Jerry Bauer

About the Author

George Smoot attended the Massachusetts Institute of Technology, where he received his B.A. in mathematics and physics in 1966 and his Ph.D. in 1970. He is a leading researcher at the Lawrence Berkeley Laboratory, and he is also a member of the Center for Particle Astrophysics and the Space Sciences Laboratory, both at the University of California, Berkeley. He is most famous for his research on the cosmic background radiation—the radiation thought to be the relic of the intense heat of the early big bang. Dr. Smoot was awarded the 2006 Nobel Prize in Physics.

WRINKLES IN TIME

WRINKLES IN TIME

Witness to the Birth of the Universe

George Smoot

and

Keay Davidson

NEW YORK • LONDON • TORONTO • SYDNEY

HARPER PERENNIAL

A hardcover edition of this book was published in 1993 by William Morrow and Company, Inc.

HarperCollins books may be purchased for educational, business, or sales promotional use. For information please write: Special Markets Department, HarperCollins Publishers, 10 East 53rd Street, New York, NY 10022.

First Avon paperback published 1994.
First Harper Perennial edition published 2007.

The Library of Congress has catalogued the hardcover edition as follows:

Smoot, George.
Wrinkles in time / George Smoot and Keay Davidson
p. cm.
1. Cosmology. II. Title.
QB981.S695 1993 93-8500
523.1—dc20 CIP

ISBN: 978-0-06-134444-2 (pbk.)
ISBN-10: 0-06-134444-3 (pbk.)

Preface

A decade after the initial publication of *Wrinkles in Time,* I am pleased to add a new preface to this edition and, particularly, to note that this book still serves its original goals of informing readers about the key background of cosmology, of telling a story about what it is like to be an adventurous scientist working to understand the universe, and also of sparking the interest of young, future scientists. All of this is still true even though many new events and observations have been made—and indeed are about to be made.

While contributing to our present understanding of the universe, the exciting discovery described in this book also promised that in the future we would learn so much more in the quest to understand our origins and future fate. Two projects are now nearing completion that may confirm our hopes. I have recently returned from Europe, where first I visited the "clean room" in which the Max Planck mission satellite is being assembled and prepared for a late summer 2008 launch. Planck will be the third-generation cosmic-background satellite—following COBE, which made the cosmic-background discovery, and WMAP (Wilkinson Microwave Anisotropy Probe), which confirmed and extended the results. The next day I flew to Geneva, where I gave a standing-room-only colloquium at CERN, the world's largest particle physics laboratory. I also toured the new LHC (Large Hadron Collider) accelerator and two giant quantum-particle detectors—ATLAS and CMS—that are in a similar state of assembly as the Planck satellite and are being readied to begin taking data in August 2008.

Since *Wrinkles in Time* was written these two projects have been traveling parallel paths. Not only did they begin at nearly the same time but it also appears they will start taking data at almost a dead heat. It is likely they will be sending out significant new results in about 2010. Once again in science we are seeing that what appear to be two distinct and separate fields—cosmology and high-energy quantum physics—are deeply related, and that each can motivate and inform the other. In the longer scale, these fields are likely to merge and overlap in a major way. I suspect these programs will yield important and interesting, perhaps startling, results that will bear directly on our understanding of the universe. *Wrinkles in Time* provides the groundwork to understand the new results and observations that are soon forthcoming. We hope many of these will be exciting—and anticipate that some will revolutionize our perspective of the world.

The adventure continues. Bon Voyage!

—George Smoot

Contents

WRINKLES IN TIME

Chapter 1

In the Beginning

I was a hidden treasure and desired to be known:
therefore I created the creation in order to be known.
—Sufi creation myth

There is something about looking at the night sky that makes a person wonder. As a child I was fortunate to live in places where the sky was easily seen at night. A vivid memory is one of riding in the backseat as my family returned home from visiting our cousins. Out the back window I watched the Moon follow us across the landscape. It seemed to follow us the way my dog trailed around as I explored our big backyard and its surrounding fields and forests. When it seemed as if it were lost behind a hill or tree, the Moon would somehow drift back into view. I wondered to my parents: Are we someplace special that the Moon is keeping watch over us? Is it we or the direction we're going? How could the Moon do the same thing for everyone in the world at the same time? Is the Moon like Santa Claus? My parents explained that the Moon is very big and very far away and that the hills and trees that got in the way were tiny in comparison—just as when you put your fingers in front of your eyes, and then you move your head over a little, you can easily see again. They then told me about the Earth and Moon, including the tides and phases of the Moon. That night my world changed. Our yard, the nearby forest, my hometown, and even the long two-hour trip to our cousins' house were but a tiny portion of a much greater world. Even more, there was reason and order—beautifully explained by neatly interlocking concepts. I could discover not only new things, like ponds and tadpoles, but I could also find out what caused things to happen, how they happened, and how things fit together. For me it was like walking into a dark museum and turning on a light. There were incredible treasures to behold.

Now, four decades later, sitting in my lab, I realize that I have been able to spend a lot of time in that museum looking for and at treasures. This is the story of the search for and eventual illumination of one such treasure, one that some have called "the Holy Grail of cosmology." It is a story that begins with humankind's first musings on the stars and our own origins and continues through centuries of observation, speculation, and experiment. It involves the very large and the very small—galactic superclusters and subatomic particles. It is a story that led me to the rain forests of Brazil and the bleak, frozen plains of Antarctica, to the romance and frustration of high-altitude balloons, the mystery of U-2 spy planes, and finally the adventure of space. It is my personal story, but also the story of many others, both historical and contemporary, who have attempted to answer the oldest and most central of mysteries: How and why did the universe begin, and what is our place in it?

Cosmology is defined as "the science of the universe." As we approach the end of the millennium, cosmology is experiencing a wonderful period of creativity, a golden age in which new observations and new theories are extending our understanding—and awe—of the universe in astonishing ways. But this current golden age can be fully understood only in light of what has gone before. Scientific knowledge is always tentative, always being refined. The history of science shows a progression of theories embraced for a time, only to be overturned or adjusted when contradicted by observation.

Western cosmology begins with the Greeks, who, two and a half thousand years ago, began to make systematic observations of the cosmos. Eventually, Aristotle's view of the cosmos emerged, a view that would prevail, in spite of minor modifications, through the Middle Ages until the Renaissance. Aristotle's was an aesthetic view of the universe, which became formalized for theology. According to Aristotle, at the instant of creation the Prime Mover (Aristotle's version of the creator) set the heavens in perfect and eternal movement, the Sun, Moon, planets, and stars fixed within eight crystalline spheres rotating about the Earth, at their center. There was no such thing as emptiness; everything was filled by the divine presence. All matter was made of the four elements: earth, water, air, and fire. A fifth essence made up the spheres, a perfect substance that could neither be destroyed nor changed into anything else—this quintessence was called the aether. The heavens were perfect and unchanging, whereas the Earth was base and subject to decay.

In the Aristotelian cosmology, motion in the heavens was circular—another sign of perfection—while on Earth, if things moved at all, they did so in straight lines. The natural state of matter was to be at rest.

There were anomalies in the Aristotelian cosmology, as observers of the heavens and of the Earth could discern. For instance, the planets appeared occasionally to change course: Mars sometimes stopped and reversed direction. Still, modified by the Alexandrine astronomer Claudius Ptolemy to take account of certain anomalies, Aristotelian cosmology persisted for two thousand years, being adopted and adapted by Christian theology along the way.

In 1514 the pope asked Polish mathematician Nicolaus Copernicus to reform the calendar. Copernicus agreed, but said the relationship of the heavenly bodies and their movement needed to be resolved. This he did, and in 1543, the year he died, published *On the Revolution of the Celestial Spheres,* a document that struck at the foundations of the Aristotelian view of cosmology, and therefore at Christian theology, which had incorporated it. The work was a product of the emerging Renaissance worldview, which placed high value on logic, mathematics, and observation. The Earth was not the center of the universe, Copernicus said; the Sun was central, and the Earth orbited the Sun as other planets do. This cosmology removed humans from the central position where they had been the focus of God's constant vigilance, and, by placing the Earth in the heavens, mixed the perfect with the imperfect. This was the beginning of the end for the Aristotelian cosmos.

Ptolemaic Picture of the Universe

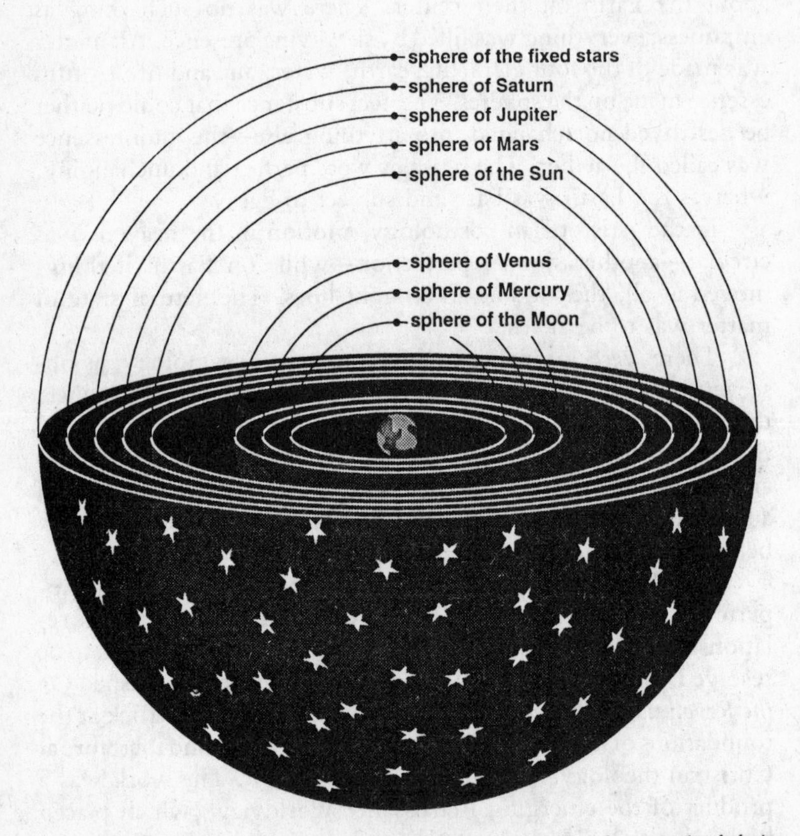

Beginning twenty-five hundred years ago, the ancient Greeks organized their observations of the world into cosmological models. One model eventually reigned supreme due to its pleasing beauty. The Egyptian astronomer Ptolemy (second century A.D.) corrected it to account for celestial observations and it was the standard cosmological model for another fourteen hundred years.
CHRISTOPHER SLYE

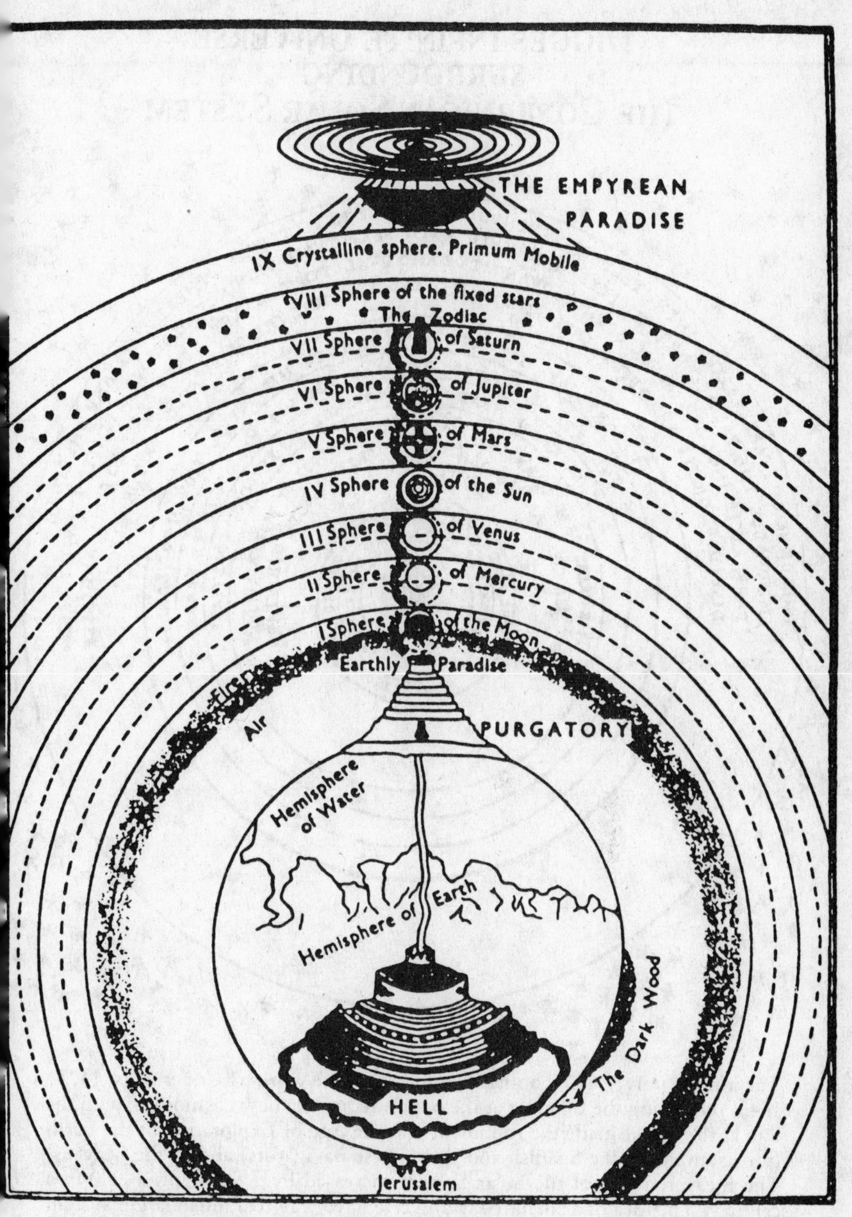

The medieval universe. Cosmology was not just the realm of science and philosophy but also included man and his pantheon. This sketch from *The Divine Comedy* by Dante (1265–1321) shows the medieval concept of the universe, including the way the Catholic Church's theology connected with Greek (Ptolemaic) cosmology.

DIGGES INFINITE UNIVERSE SURROUNDING THE COPERNICAN SOLAR SYSTEM

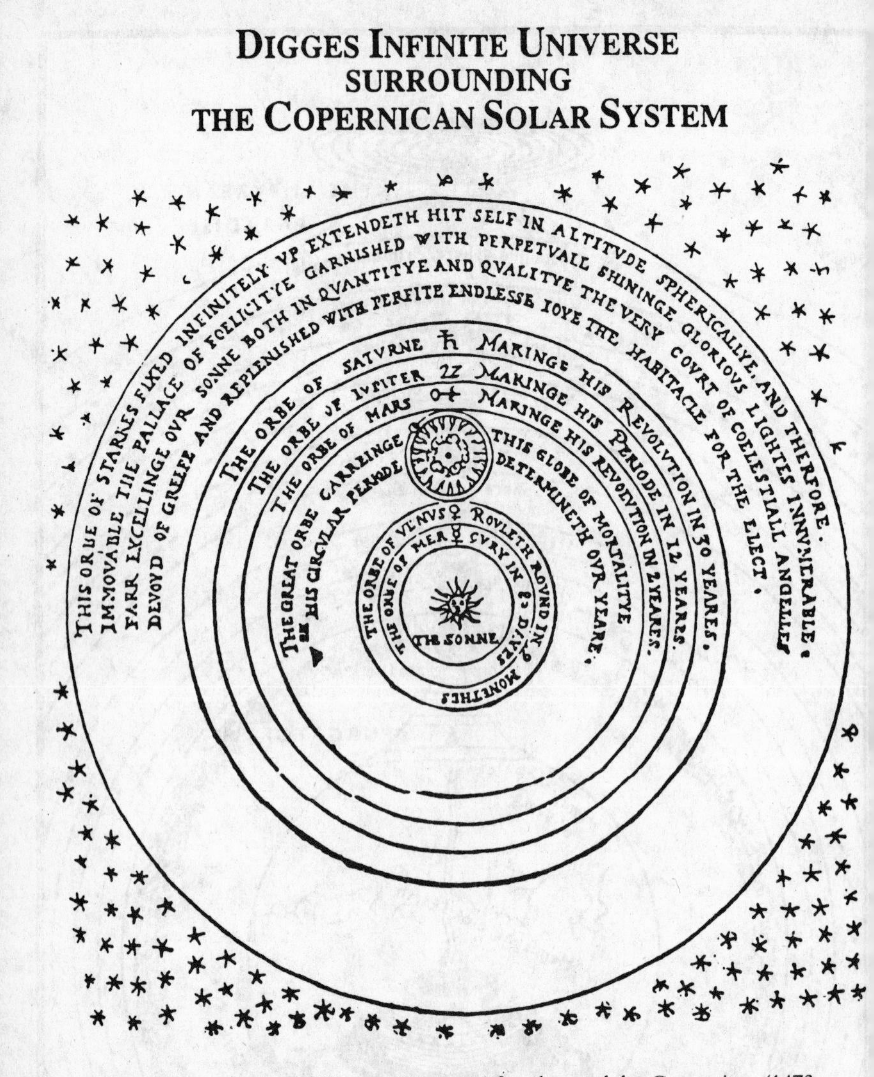

The infinite universe of Thomas Digges. After the work by Copernicus (1473–1543) reforming the calendar at the pope's request, a new cosmology with the Sun at the center made the rounds. It was the Age of Exploration of the Earth (for example by the Spanish and Portuguese navigators) and of the heavens. The measured size of the Solar System increased by a factor of 10,000 in a century. The idea of a small, cozy universe gave way to a much larger system in 1576 when Thomas Digges (1543–95) published this figure of the Copernican system, combined with an outer orb of stars extending infinitely.

Big Bang:
Expanding and Evolving Universe

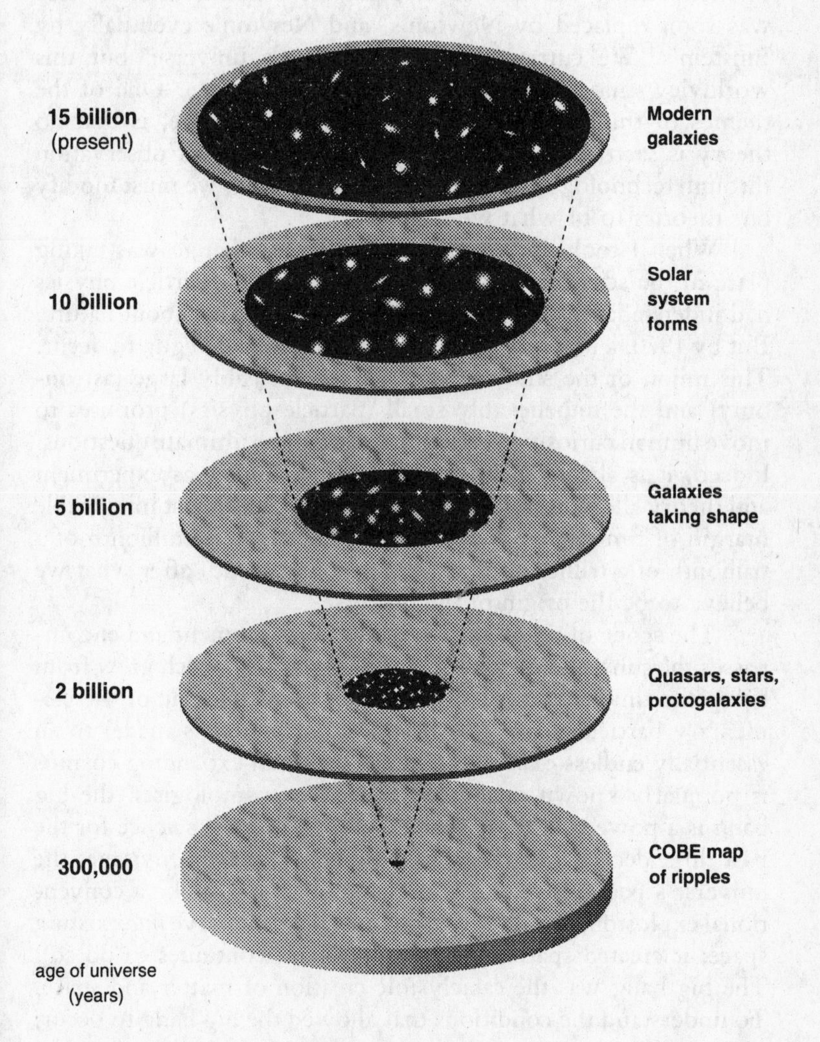

Big bang cosmology holds that the universe is expanding and evolving. When one looks back in time, the universe is more dense and hotter and its contents younger. At the very beginning, there are only seeds. CHRISTOPHER SLYE

Over the past four centuries a series of astronomical observations and experiments has radically altered our view of the universe. Just as Aristotle's geocentric universe was replaced by the heliocentric universe of Copernicus, Copernicus's universe was soon replaced by Newton's, and Newton's eventually by Einstein's. We currently live in Einstein's universe, but this worldview may also someday become inadequate. One of the themes of this book, and of the history of science, is that no theory is sacrosanct. As we expand our powers of observation through technology and experimental ingenuity, we must modify our theories to fit what we see.

When I took up cosmology in 1970, a change was taking place in the science. In the past, astronomy and particle physics had independently pursued fundamental questions about nature. But by 1970, a union of these two disciplines had begun to occur. This union of the study of the incomprehensibly large (astronomy) and the unbelievably small (particle physics) promises to move human curiosity closer to answering the ultimate questions. Indeed, it is already delivering on that promise, as experiment and theory allow us to look back toward the slimmest imaginable margin of time, some 10^{-42} seconds (that is, one millionth of a trillionth of a trillionth of a trillionth of a second) after what we believe to be the origin of the universe.

The scope of cosmology begins at that moment and encompasses the subsequent evolution of our cosmos, which grew from being the tiniest fraction of the size of a proton (one of the elementary particles from which all known matter is made) to an essentially endless expanse. This theory of an expanding cosmos is popularly known as the big bang. To cosmologists, the big bang is a powerful theory that has dominated the science for the past three decades. As the words imply, the theory envisages the universe's beginning with a mighty eruption. Unlike a conventional explosion, however, the big bang did not move *into* existing space; it created space as it expanded (and continues to do so). The big bang was the cataclysmic creation of matter and space. To understand the conditions that allowed the big bang to occur, we must abandon our commonsense notion of matter and energy and time and space as separate. The universe at the moment of creation existed under very different conditions and probably operated according to different laws than it does today. Reality in cosmology sometimes evades our comprehension.

Though the pedigree of big bang cosmology reaches back to an idea developed between 1927 and 1933 by Georges-Henri Lemaître, a Belgian priest, it wasn't until 1964 that the theory emerged as the dominant explanation of how the universe came to be the way it is. In that year two American radio astronomers discovered what appeared to be the dim afterglow of the ancient cataclysmic event. That afterglow, an all-pervasive hum of radiation with a temperature equivalent of a little more than 3 degrees Kelvin (three degrees above absolute zero), is known as cosmic background radiation, and it provides us with a faded snapshot of the universe as it was some three hundred thousand years after the big bang. It is within the background radiation that my colleagues and I hoped to discover our wrinkles in time, the Holy Grail of cosmology.

One of the greatest challenges to the big bang theory has been to explain how matter is distributed through the ever-expanding space of the cosmos. It is possible to imagine that all matter would have been scattered evenly through space, making the universe a homogeneous cloud of gas, with an average density of about one hydrogen atom every ten cubic meters. (For comparison, the air we breathe contains 3×10^{25} atoms per cubic meter, mainly atoms of nitrogen, oxygen, and carbon, all of which are bulkier atoms than hydrogen.) Had the universe today, some 15 billion years after it formed, been a virtually unending cloud of gas, the night sky would be unrelentingly black, and we would not be here to observe it. However, we know from our very existence that something in the evolution of the universe caused matter to condense, to form stars and planets—and, ultimately, life (not just life on Earth but, with a probability approaching 100 percent, on millions of other planets, too, including some in our own Milky Way).

Again, it is possible to imagine that stars, like our own Sun with its orbiting planets, could have been distributed evenly throughout the universe, a uniform cloud of countless billions of points of light in the night sky. But, again, we know from our own experience that that is not the case. Our sun is but one of a

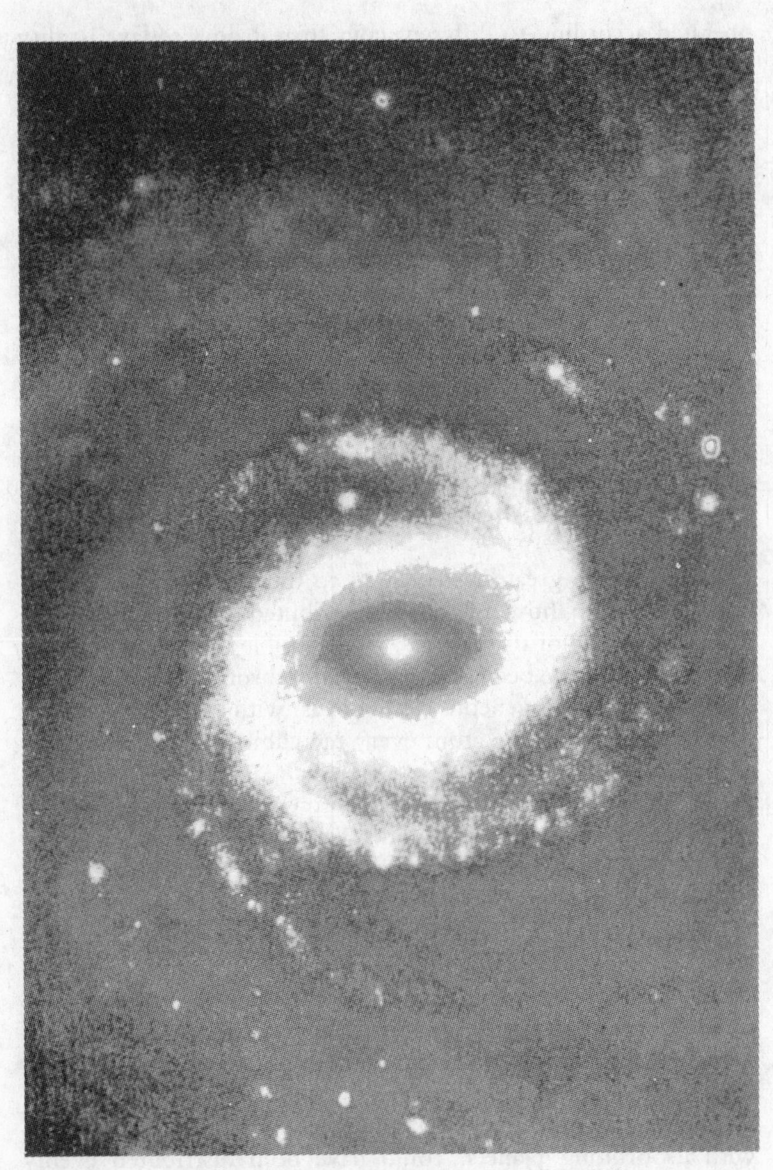

A photograph of a spiral galaxy, NGC 1232, 65 million light-years away. Visible matter is clumped in stars and the stars are organized in giant systems called galaxies. If we could see our own Milky Way, this is probably what it would look like. CALIFORNIA ASSOCIATION FOR RESEARCH IN ASTRONOMY hundred million similar stars in a huge, disk-shaped, rotating spiral galaxy, the Milky Way, seen as a wispy band across the night sky. To all intents and purposes, all stars are members of such galaxies. Matter is therefore clumped together not only as stars but also as collections of stars, or galaxies.

Again, it is possible to imagine that all galaxies, once they had formed from the condensation of matter into a community of stars, could have been distributed evenly throughout the universe, a uniform cloud of blurred spirals in the night sky. A major

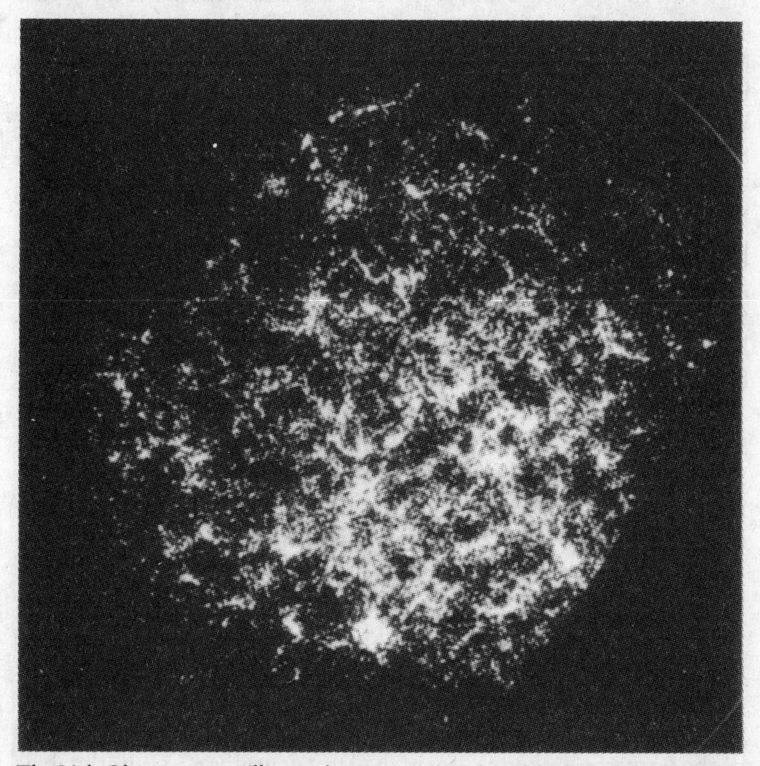

The Lick Observatory million-galaxy survey. A plot of the location of a million galaxies covering one hemisphere of the sky. Note how the galaxies are not evenly or randomly distributed, but are organized. They are in clusters and groups of clusters. There are spherical voids with galaxies on the surface and clusters at the intersections of the void bubbles. The distribution is like foam. Matter is grouped on a scale from stars up to the largest size observed. EDWARD SHAYA, JAMES PEEBLES, AND R. BRENT TULLY discovery of recent cosmology is that this, too, is not the case. Galaxies are often collected together, not only as clusters of thousands of galaxies, but as even larger entities known as superclusters, and structures larger yet, some many millions of light-years in extent. In other words, matter in the universe is highly structured. A useful image of the universe is a foam composed of soap bubbles, in which the walls of the bubbles represent concentrations of galaxies and their interiors represent vast empty volumes of space.

But the structure and formation of visible matter is only part of the conundrum for the modern cosmologist. Go out tonight and, if you are blessed with a clear sky and little extraneous light, look deep into the heavens. If you use binoculars or a telescope you will see a night sky ablaze, such as Galileo saw it four centuries ago, holding millions of stars and galaxies, the stuff of creation. This is what we usually think of when we talk about the universe. However, it is what you are *not* seeing that is of increasing importance to theorists. If modern cosmology is correct, the shining stars in the dark night sky represent less than 1 percent of the stuff of creation. Most of matter created during the big bang may be completely alien to us: invisible to our eyes and quite beyond our physical experience.

This giant new cosmological puzzle relates to the central search of cosmology these past three decades. The discovery in 1964 of cosmic background radiation appeared to confirm the reality of the big bang. But it left unanswered a key issue: How did the big bang lead to the formation of stars, galaxies, galactic clusters, and so on, by the condensation of the stuff of creation? If the big bang happened, clues to the formation of the structures we see in today's universe should be evident in the earliest remnants of the fury of creation. The clues should be evident in the cosmic background radiation.

To its discoverers, the background radiation from all regions of the universe looked uniform, a picture showing a smooth fabric of space and energy. But in order for structures to condense from the products of the big bang, the smooth fabric must have borne tiny wrinkles, fluctuations in temperature caused by areas of higher density. According to the big bang theory, matter (familiar

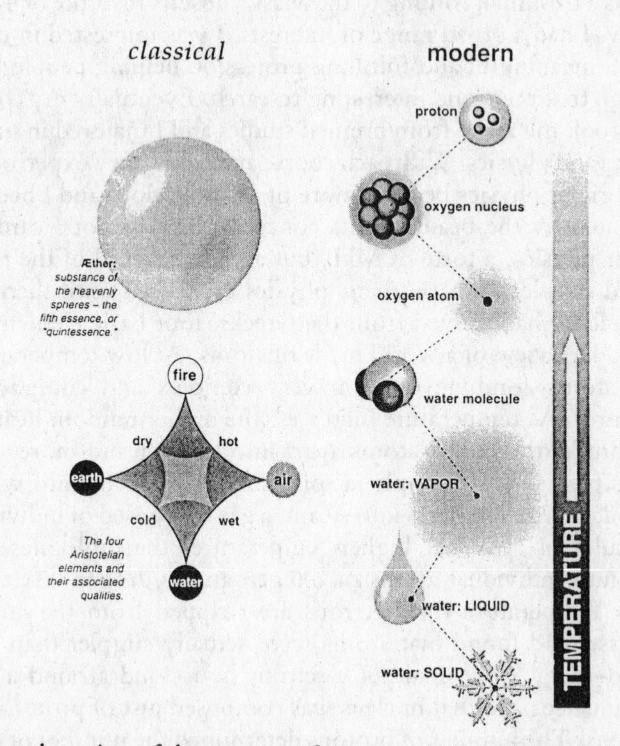

Our modern view of the structure of matter makes no distinction between celestial and terrestrial matter, as did that of the ancient Greeks. The perfect crystalline aether of the celestial spheres is supplanted by the notion of matter and physical laws that are the same here and there. Our current concept is that matter is made from simple building blocks put together into more and more complicated structures. Each level of complication is more delicately held together. Raising the energy of interactions (i.e., the temperature) breaks apart the matter into more basic building blocks. This has been the direction of progress for physics. The universe is built in the opposite direction—from simple to complex. CHRISTOPHER SLYE

and unfamiliar) could have condensed and subsequently formed galactic structures in such areas through gravity. These wrinkles—we can also call them cosmic seeds, from which the galaxies grew—must have been present, otherwise modern cosmology, and specifically the big bang theory, would be in serious trouble.

* * *

As a freshman coming to the Massachusetts Institute of Technology, I had a broad range of interests. I was interested in medicine, a meaningful and fulfilling profession helping people both through treatment and interesting research. Eventually my course work took me away from premed studies and I majored in mathematics and physics. With each course and laboratory experiment, the fabric of physics became more of a whole cloth and I became captivated by the beauty of its concepts. As a senior I came to nuclear physics, a forte of MIT, but a giant reversal of the trend toward simpler more aesthetic physics I had seen until then.

Science had moved from the Greeks' four basic elements to the modern view of a world made of atoms. At low temperatures these atoms combine to form very complex and complicated structures. As temperature increases, the higher random heat energy breaks the linked atoms apart into simpler and more symmetric pieces. For example, a solid snowflake melts into water, a liquid, which vaporizes into steam, a gas composed of individual molecules of H_2O. At higher temperatures the molecules split apart into individual atoms of oxygen and hydrogen. At a still higher temperature the electrons are stripped from the atoms. Scientists had found that atoms were actually simpler than they looked—they were made of electrons in a cloud around a tiny dense nucleus and that nucleus was composed just of protons and neutrons. The number of protons determined the number of electrons in the atomic cloud and thus the chemical properties of the atom. The electrons on the outside of the atomic cloud were the handles by which atoms could join together to make substances. The electrons were clearly simple—very lightweight, pointlike particles, all identical, and carrying a single unit of electric charge.

This made a wonderfully simple picture: All of matter is made by combining in different ways just three simple particles, the electron, proton, and neutron. These composites, in turn, combine to make even more complex structures. Unfortunately, nuclear physicists found out that the proton and neutron are not simple, like the electron. In high-energy collisions, new particles were being discovered by the handful. The force holding the neutrons and protons in the nucleus (the strong nuclear force) turned out to be not nearly as neat and simple as the electrical force holding electrons to atoms, or the gravity holding us on the Earth. At high energies, physics seemed to have turned messy. The strong force was very complicated. This was a reversal of the pattern that made physics so beautiful and fundamental. Ideally, physics should reduce the number of things that are separate and that have to be remembered and explained. Instead, the interactions were now more complicated and there was a plethora of particles with no known use.

This also gave me pause about the big bang model. If elemental nuclei were so hard to understand, how could anyone hope to understand Georges-Henri Lemaître's primordial atom with its nucleus as big as the Solar System? At nuclear densities the universe would be very complicated. How could we find out how the universe began?

Fortunately, at this time Professor David Frisch recruited me to work with his group for my senior thesis project. His group had set up a detector at the big accelerator at Brookhaven National Laboratory on Long Island. They had gathered reams of data showing the debris resulting from collisions of the accelerator's high-energy protons with deuterium nuclei in the detector. (A deuterium nucleus contains a single proton and neutron and makes up heavy hydrogen.) In this debris were traces of many particles. Most of those particles were so ephemeral that, traveling at nearly the speed of light, they could travel only a hairsbreadth or less before their lives were over and they decayed into other particles. He asked me to investigate some of the data. I did, finding that about one third of the time a mysterious particle, the eta-naught, decayed into three lighter particles, pi-naughts, which immediately decayed into two gamma rays (energetic light quanta) each. It was a difficult task, because I had to identify six gamma rays in all the debris and determine if they came from the decay cascade. However, it was rewarding to scan the pictures, measure the tracks, calculate the properties and probabilities, and finally learn something about these elusive, exotic things.

Frisch liked my results and invited me to work with another newly graduated student, Don Fox, to create an experiment to find even more short-lived particles that might decay into K-naught particles. K-naughts behave strangely, having a mysteriously long lifetime and thus traveling macroscopic distances before decaying. Soon we were at work. In a couple of years I found myself working on a doctoral thesis in another experiment testing how well the decays of such strange particles follow certain rules—in particular, to determine if a change in strangeness of a particle is always accompanied by a matching change in electric charge.

The particle physics community viewed this as a fundamental experiment, and soon we—Dave Frisch, Orrin Fackler, Jim Martin, Lauren Sompayrac, and I—were in competition with four other groups. We worked with giant magnets, particle counters, and track detectors at the Brookhaven atom smasher, and with computers back in our labs and offices analyzing the data. While we were deeply involved in this experiment, eager to do the best job, Jerome Friedman and Henry Kendall's group, in the next office complex down the hall, were running an experiment to measure the size and shape of the proton. They used the Stanford electron accelerator because electrons are simple pointlike probes of the complicated protons. Stanford accelerator theorist James Bjorken showed that their experiment demonstrated that the proton is not fundamental but is in fact made of smaller pieces—which they called partons. They had found that protons and neutrons are made of simpler pointlike particles now called quarks. The interaction of these quarks gets weaker and simpler the more closely packed the quarks are and the higher the energy of the quarks. All the extra particles that physicists had found were simply unstable combinations of quarks that quickly split into more stable combinations.

This was stupendous. Now I felt physics was back on track. At higher energies, things get less complicated and more symmetric. The big bang looked much more tractable. Physics became simpler and easier the closer one got to the beginning of the universe. It was now easy for me to imagine everything in the universe compressed into a region smaller than a proton. When one reaches Lemaître's primordial nucleus, just keep pushing. The protons and neutrons dissolve into a soup of quarks. If the quarks are really pointlike, or at least very, very tiny, then it is no problem to push them into a region the size of a proton. With a great many quarks packed together, they resist less than when there are only three in the volume of a proton. This compression does, however, require an unimaginably high temperature.

These realizations, and the fact that experimental particle physics was becoming extremely crowded (leaving less opportunity for a single individual to make an impact), led me to make my switch to cosmology in 1970. Big bang theory was already dominant when I entered the field, but the process of the big bang—its cataclysmic event, the creation of all matter, and the formation of the galaxies—seemed almost mystical to me. I remember on more than one occasion contemplating the night sky and musing that the big bang seemed as hard to swallow as the image of Atlas supporting the world on his shoulders, causing earthquakes as he shifted position to gain temporary comfort—perhaps even harder! Twenty years later, it can still seem mystical to me, not because it is unscientific, but because it represents something so important to the human psyche.

Others feel this, too, as evidenced by the frequent appearance of newspaper stories about the theory—that some new piece of evidence supports it or threatens it, or that the absence of evidence leaves it unsupported. So much attention to a scientific theory by society at large speaks of the mythical force of the theory. People know it is science, but want it to be more. For instance, at the 1993 annual meeting of the American Association for the Advancement of Science, held in Boston, in a session called "The Theological Significance of Big Bang Cosmology," scientists and theologians sought connections between the fundamental science of the big bang, as described by current theory, and the Christian story of creation. There is no doubt that a parallel exists between the big bang as an event and the Christian notion of creation from nothing. (Creation from nothing does not appear in the Bible itself, but was formulated later to exclude the Gnostic teaching that matter is evil, the work of a lesser being, not the work of God.) It is of no small significance that in 1951 Pope Pius XII invoked big bang theory and observational evidence that supports it: "Scientists are beginning to find the fingers of God in the creation of the universe."

Man has always been obsessed by the search for his own origins. Creation myths are ubiquitous, and in the ancient world images of the cosmos were often central to all aspects of life—religious, political, and military. It is therefore not surprising that the big bang, even in this modern secular world, often takes on the dimensions of myth.

As science, the big bang is a powerful theory for explaining the origin and the evolution of the universe. But our desire to understand the universe reaches far deeper than the history of science and its rational methods. As Joseph Campbell, the world's foremost interpreter of mythology, has said: "What we humans are looking for in a creation story is a way of experiencing the world that will open to us the transcendent, that informs us and at the same time forms ourselves within it. That is what people want. This is what the soul asks for." Society hungers for both science and mythology, and the big bang theory is where the two mingle most intimately.

In the following pages I recount the journeys of others in their quests for insight into what makes the universe the way it is—the heroic endeavors that helped develop the remarkable creation story we call the big bang theory. And I recount my colleagues' and my own adventure, the search for the wrinkles in time, those distant echoes of the early formation of the galaxies.

Chapter 2

The Dark Night Sky

In the winter of 1984 I worked in Rome for a week. I was attending a scientific workshop on the early universe and a meeting with R. Mandolesi, G. Sironi, L. Danese, and G. De Zotti of the Universities of Bologna, Milan, and Padua. For four years we had been collaborating to measure the intensity of the cosmic background radiation, a project that had become a major part of my career as a cosmologist. I had visited Italy a couple of times previously, but had not yet seen the Leaning Tower of Pisa. This time I vowed to make the trip.

Impatient for the conference to end, I rented a car on a Friday afternoon and set off to drive the two hundred miles northeast of Rome to Pisa, an old city on the coast of Tuscany. Despite driving as fast as I dared on Italian roads, I arrived as dusk was falling and feared I would be too late to enter the tower. At a gas station on the south side of this impressive walled city I sought directions—in broken Italian—to the "Leaning Tower of Pisa." "*Ah, Piazza dei Miracoli!*" replied the attendant, in evident reverence. Assuming we were both talking about the same thing, I followed his instructions to circle the city outside the wall and enter by the northwest gate.

I hurriedly parked and rushed through the gate, even though by then I knew that I would only be able to see the tower, not go in. There it was, the piazza—the cathedral was directly in front of me, the Leaning Tower looming just behind the cathedral, and between them rose a full moon, its light shining on the white marble of the tower. It was one of those rare times when high expectations are exceeded by reality. The architecture, the dark green grass, and the white marble gleaming in the moonlight made a breathtaking scene I will never forget. I knew then why it is called the *Piazza dei Miracoli*—the Place of Wonders.

No one needs a reason to visit Pisa, beyond its splendid and unusual architecture. But for me there was another reason: Legend has it that Galileo Galilei (1564–1642) performed an experiment at the Leaning Tower that essentially constitutes the foundation of modern physics and modern cosmology. From the top of the tower he is said to have dropped two objects of differing mass, to see if they would hit the ground simultaneously. They did, thus demonstrating that all falling objects accelerate at precisely the same rate, regardless of mass.* In addition to being the first experimental physicist, one who took the study of motion from abstract philosophy into concrete science, Galileo was also the first astronomer to turn a telescope to the sky. The presence of the rising Moon when I arrived at the piazza was therefore doubly appropriate: We understand its motion thanks to Newtonian physics that evolved from Galileo's experiments; and we understand its terrain thanks to his pioneering observations with a telescope.

As a child I read of Galileo's tower experiment and was delighted by its simplicity and drama. Galileo became a mythic figure for me, a true hero in the long journey of science to understand the world. I understand that some historians consider the story to be apocryphal, but I believe it. Galileo already knew such an experiment would work, because he had tested its scientific principle by rolling spheres of various masses down an inclined plane. These objects descended the inclined plane more slowly than spheres dropped from a tower, hence their motions were easier to time. It is easy to imagine Galileo's scientific showmanship in demonstrating what he already knew to be true, by

*Galileo's principle of falling bodies was demonstrated during one of the Apollo flights to the Moon, which is airless. An astronaut dropped a feather and a heavier object while a global TV audience watched. They hit the lunar soil at precisely the same instant.

Galileo Galilei, whose insights were the foundation of Newtonian physics.
YERKES OBSERVATORY

dropping spheres from the fabled tower before an incredulous audience. It would have been a powerful visual argument for a fundamental scientific law, and a blow to his Aristotelian colleagues who taught at the university. Not long after this Galileo had to leave stuffy Pisa for the more open university in Padua.

The next morning I climbed the tower, and, standing where Galileo surely stood, I could feel that such an experiment would have been irresistible. I went from the tower to the cathedral, Pisa's *duomo,* where an attendant came up to me and pulled me over to the nave, pointed upward, and said: "*Lanterna di Galileo.*" Hanging three stories from the soaring ceiling was the lamp that Galileo had watched while attending cathedral services in the 1580s. It is said that it was Galileo's responsibility to make sure the lamp was burning, and so he kept a close watch on it. But he did more than monitor the flame. He was a medical student at the time, and he used the regular beat of his pulse to time the swings of the lantern as it was nudged by air currents. Common sense suggests that the wider swings would take longer. But Galileo saw that all swings, regardless of amplitude, took exactly the same time: Galileo had discovered the principle of the pendulum. Soon, he was doing more experiments and learning about motion and inertia.

Until the late sixteenth century, philosophy had developed abstract, idealized concepts of nature divorced from physical reality. This was insufficient for Galileo, who wanted to learn the laws of nature by doing experiments and by making observations. He made a telescope through which he saw the phases of the planet Venus (which provided direct proof of the Copernican theory that the planets revolve around the Sun); he saw satellites orbiting Jupiter (another, less direct evidence for Copernicanism, in that not all heavenly bodies orbited the Earth and the Jovian system resembled a mini–solar system); he spotted countless stars in the Milky Way (evidence that the universe was far larger than then believed); and he discovered craters and mountains on the Moon, plus broad, flat areas he assumed (wrongly) were seas. These Earth-like features among the heavenly bodies—plus spots he detected on the Sun—proved the heavens weren't as "perfect" as contemporary theology had assumed. Perhaps the celestial "aether" was a myth; perhaps Earth and sky consisted of the same "stuff"—stuff that obeyed the same laws of motion. (These insights eventually allowed Newton to perceive the link between a falling apple and the orbiting Moon.) Thus collapsed the medieval barriers between heaven and Earth, between aether and matter. In a sense, physics and astronomy became one subject under Galileo's genius. It was the grandest marriage of two physical sciences until the late twentieth century, when cosmology and quantum particle physics began to merge.

At first, church officials tolerated Galileo's writings, despite their challenge to established dogma about the universe. But when he publicized his ideas and observations in books (such as *The Starry Messenger*) written in Italian (not the more academic Latin) for a mass audience, the cardinals grew edgy. In 1633 the Inquisition forced him, on his knees, to recant his views. As he rose to leave, he allegedly muttered, "And yet it [Earth] moves." House arrest followed, then blindness, then death in 1642.* Nevertheless, he was eventually buried in the church of Santa Croce in Florence along with other notables such as Machiavelli, Michelangelo, Dante, and Rossini. Santa Croce is one of my favorite places for pilgrimage. It also holds a lot of fabulous art, sculpture, and architecture (including Brunelleschi's Pazzi Chapel)—but most important, it is only a couple of blocks from Vivoli's gelateria, which serves the best ice cream in the whole world.

Galileo's experiments, with the observations of Danish astronomer Tycho Brahe (1546–1601) and the theoretical work of the German astronomer Johannes Kepler (1571–1630), paved the path to Newtonian physics. That is where most of the science described in this book begins. Newton's equations are used in countless ways, from modeling the formation of galactic superclusters to plotting the orbits of satellites. If Einstein was the father of modern cosmology, and Newton its grandfather, then Galileo was surely its great-grandfather. Each of these three geniuses in turn wrought the intellectual revolution of his time, dramatically altering our view of the universe. The perceived universe of Galileo's time was finite and static. With Newton's laws of motion and gravity, and with Einstein's theories of special and general relativity, the universe would come to be seen as infinite and dynamic.

. . . what hinders the fix'd Stars from falling upon one another?
—Isaac Newton
Opticks *(1704)*

The idea of an infinite cosmos was shocking to established orthodoxy, which was solidly grounded in theology. The old

*In 1992, three hundred and fifty years later, the Vatican issued an apology for the church's treatment of the Florentine stargazer. Now the church maintains astronomers and other scientists to advise it about the physical world. At present it is hard to imagine scientific societies having theologians and others to advise them on the spiritual world. Scientific societies do maintain lawyers and occasionally ethics committees to advise them. As science matures and finds its domain we are likely to see more interaction of this type.

Christian cosmos had been tiny and cozy; humans were at the center of all creation, literally watched over by God and the angels. But by the late seventeenth century, the possibility of an infinite universe had gained wide attention in intellectual circles. Copernicus's cosmology provided a solid scientific reason for suspecting the cosmos was much bigger than ever dreamed, if not infinite.

The possibility of a vast, perhaps infinite, universe was discussed in 1576 by the English author Thomas Digges. He described the "orb of stars fixed infinitely up . . . perpetual shining glorious lights innumerable far excelling our sun both in quantity and quality." Better remembered is the courageous, abrasive Italian monk Giordano Bruno, who insisted there were "innumerable suns, and an infinite number of earths revolve around these suns." Bruno also expressed radical political views, an unwise move in those days of Reformation and Counter-Reformation. In 1600 he was burned at the stake.

If Bruno was right, then Earth was but one planet orbiting but one star among innumerable stars. We've all heard legends of Pacific islanders who became disoriented when they learned there are lands beyond their own shores. Many Europeans felt the same way in the late seventeenth century when they looked at the Milky Way, that newfound sea of suns, and imagined countless worlds swirling around those suns. The French mathematician Blaise Pascal expressed the sentiments of many when he wrote: "The eternal silence of those infinite spaces strikes me with terror."

The climax of the Copernican revolution, and with it the eventual acceptance of an infinite universe, came with the publication of Newton's *Principia*. Isaac Newton (1642–1727), who was born the year Galileo died, has been the most influential scientist since Copernicus, and is rivaled only by Einstein. Newton's laws of motion and of universal gravitation have had a profound impact on science and are among the most far-reaching generalizations formulated by the human mind. In his *Principia,* Newton showed how the new cosmology made physical sense in terms of the laws of motion. He had derived three laws of motion from Galileo's experimental data and Kepler's theoretical work on planetary motions.

Isaac Newton, who unwittingly "discovered" the possibility of an expanding universe by proposing his theory of gravitation and his laws of motion. YERKES OBSERVATORY

Newton's laws of motion are these: (1) a moving body will continue in uniform motion until something acts to divert it (this is the principle of inertia); (2) any change in motion of a moving object depends on the force applied to the object, divided by the mass of the object; and (3) for every action there is an equal and opposite reaction. Newton's law of gravitation is this: The force of gravity between two bodies is proportional to the product of their masses, and gravity diminishes according to an inverse square of the distance between them. For example, the gravitational pull between two masses becomes one fourth as strong when their distance of separation is doubled, one ninth as strong at three times the distance, and so on.

These laws apply to celestial and terrestrial bodies alike; they explain why the orbiting Moon is literally like a falling apple, in that the Moon is falling "around" Earth. With these laws Newton provided the first correct explanation for planetary orbits. In principle, his laws allow the exact position of planets to be predicted centuries in advance. Eventually astronomers showed that Newton's laws applied to the universe as a whole. Comets such as Halley's* travel around the Sun on highly elliptical orbits that can be predicted with Newtonian equations; binary stars orbit each other according to Newtonian laws; and brand-new planets can be discovered (as was Uranus) by using Newtonian laws to detect their gravitational pull on known planets.

But Newton's theory posed a paradox. The nature of this paradox was pointed out in 1692, in a letter he received from the Reverend Richard Bentley, a brilliant and tempestuous scholar at Cambridge University. The correspondence between the two men dealt with the effects of gravity on an infinite universe and mark the beginning of what we might call dynamic cosmology: that is, the study of the universe as an evolving entity. Their exchanges were the first, faltering steps on the long road to the big bang theory. If the cosmos is infinite, wrote Bentley; and if, as Newton's theory held, gravity pervades the cosmos (diminishing by the inverse square of the distance but never disappearing altogether); then simple calculation shows that every part of the universe should be subject to an infinite gravitational pull. Hence, all the stars should collapse into a giant fireball.

In his initial response to Bentley, dated December 10, 1692, Newton acknowledged the possibility that "if the matter of our Sun & Planets and the matter of the Universe was evenly scattered throughout all the heavens, & every particle had an innate gravity towards all the rest & the whole space through which this matter was scattered was but finite: the matter on the outside of this space would by its gravity tend toward all the matter on the inside & by consequence fall down to the middle of the whole space & there compose one great spherical mass."

The universe might be saved, Newton added, if the stars were uniformly distributed across infinite space. Each star, he speculated, is separated by precisely the same distance, as neatly

*In 1705 the English astronomer Edmund Halley noticed that comets from 1531, 1607, and 1682 had approximately the same orbit. He suggested a single comet was involved. Using the new method of his friend Isaac Newton, he predicted its orbit and return in 1758. Sure enough, it reappeared sixteen years after his death and now bears his name.

spaced as squares on an infinite chessboard. If so, then each star would feel each other's gravitational pull equally in all directions out to infinity. Any tendency to fall in one direction would be balanced by an equal tendency to fall in the opposite direction. As a result there would be no movement of stars and no cosmic collapse.

But Newton quickly realized this "solution" was extremely unstable. He wrote to Bentley again a month later, indicating that the slightest movement of a single star would trigger gravitational perturbations throughout the system. Hence the entire system would collapse into a single heap under the influence of gravity or, alternatively, different parts of it would accumulate in countless individual heaps. Still, Newton added, the scheme might work if a "divine power" intervened to ensure that the stars "would continue in that posture [spaced at equal distances] without motion for ever." Newton was clearly feeling desperate. Already, in those distant days, it was becoming unfashionable to invoke divine power to patch up embarrassing holes in scientific theories.

Newton and Bentley had both assumed a static universe; it didn't occur to them that the stars might change position over time. Had they assumed a nonstatic universe, then they could have seen that motion of the stars could have counteracted stellar collapse. But their failure to do so is understandable. At that time, no one had detected such motion. The concept of a finite universe had been abandoned in favor of an infinite universe through the insights of Newton and others. The concept of a static cosmos would, however, persist well into the early twentieth century.

The concept of an infinite universe does not explain the one fact that is obvious to everyone: The sky is dark at night. This may sound like a trivial observation, but it is not. The seventeenth-century astronomer Kepler was one of the first to recognize the darkness of the night as a mystery. If the number of stars is infinite and they are uniformly distributed, then they would cover every part of the night sky, with no gaps between them. In that case the heavens would glow like a fireball and bake the Earth. Life would be impossible. Hence, Kepler argued, the universe is finite, not infinite.

Nocturnal darkness also puzzled the German physician-astronomer Heinrich W. M. Olbers in the 1820's. By day he practiced ophthalmology and fought epidemics; by night he watched for comets and asteroids from the top floor of his home in Bremen. Unlike Kepler, Olbers believed the cosmos was infinite, and he proposed a way to reconcile this belief with the dark night sky: Space is pervaded by clouds of interstellar matter that intercept much starlight, dimming it, just as an umbrella shades the sun. He made this interesting suggestion *before* the existence of interstellar clouds was firmly established. However, his invocation of such clouds to explain the paradox isn't valid; in an infinite cosmos, starlight would eventually heat the clouds until they glowed, making the night sky burn as brightly as Kepler had conjectured. Nonetheless, the night-sky puzzle has been known ever since (somewhat unfairly to Kepler and other predecessors) as "Olbers's paradox."

A true explanation was uncovered more than a century ago, and not by a scientist but by a poet—Edgar Allan Poe (1809–49). Most people know of Poe's horror tales, the belles lettres of Gothic terror; and of his messy personal life, of the drinking that killed him when he was forty. But few know of his serious interest in science, especially astronomy, and his fascination with the French astronomer Pierre-Simon de Laplace's nebular hypothesis, according to which the Solar System evolved from a primordial cloud of dust and gas. The hypothesis "is by far too beautiful *not* to possess Truth as its essentiality," Poe wrote. It inspired him to write a cosmological essay, *Eureka: A Prose Poem*.

According to Poe, Olbers's paradox is resolved because "[The] distance of the invisible background [is] so immense that no ray from it has yet been able to reach us at all." What Poe had stumbled upon was the fact that the universe is not infinitely old, but had a beginning in time (a point we now regard as the big bang). In fact, the universe is so young that light from the more distant stars is still speeding toward us but hasn't yet reached us. When we gaze at the dark night sky, we look back to a primeval epoch, before the earliest stars had formed.

Eureka got mixed reviews. "Hyperbolic nonsense," scoffed one newspaper. Nevertheless, Poe remained almost frantically aware of the importance of his insights, and showed the manu-

Dark Night Sky
(Olbers's Paradox)

IF the universe is static, infinite, and uniform…

THEN every line of sight must end on the surface of a star.

WHY aren't we fried?

Solutions:

- Universe finite in time
- Universe expands

The night sky is dark because the universe isn't infinitely old; hence, light from the more distant stars hasn't reached us yet. CHRISTOPHER SLYE script of *Eureka* to the publisher George Putnam in New York City. Putnam later remembered Poe's "tremor of excitement . . . [He claimed *Eureka*] was of momentous interest. Newton's discovery of gravitation was a mere incident compared to the discoveries revealed in this book. It would at once command such universal and immense attention that the publisher might give up all other enterprises, and make this one book the business of his lifetime. An edition of fifty thousand copies might be sufficient to begin with. . . ." Putnam was "really impressed—but not overcome," and printed five hundred copies instead. Six months later, Poe was dead.

The First World War (1914–18), the historian Barbara Tuchman has written, was the chasm that forever separated the old world from the new. Much of civilization—the Edwardian era of empires and colonies, monarchies and aristocracies—was transformed during that conflict. So was much of the intellectual world, the world of art, literature, and science. Cosmology was no exception.

Until the late 1910's, humans were as ignorant of cosmic origins as they had ever been. Those who didn't take Genesis literally had no reason to believe there had been a beginning. The origin of the Solar System was a contentious topic, but the origin of the entire cosmos was an altogether different matter: It was rarely, if ever, discussed in scientific circles. In the astronomical journals of the day there was much discussion about the nature of the nebulae, the 1910 return of Halley's Comet, the evolution of stars, the Martian "canals," the Balmer series in stellar spectra, the search for a ninth planet—but hardly a word about cosmic origins.

As was true for the other major intellectual changes of the time, there were clues that a revolution was coming. For the most part, the import of these clues was not appreciated, so that the new ideas, when they finally coalesced, came as a breathtaking shock.

Between 1905 and the middle of the next decade, Albert Einstein offered his "special" and "general" theories of relativity and changed forever the way scientists view the universe. Special relativity showed that space and time, mass and energy were, respectively, opposite sides of the same coins. General relativity revised Newtonian gravity by showing a connection between what Einstein called mass-energy and space-time. Where Newton thought of gravity as a force acting between objects, Einstein viewed it as the effect of mass on the geometry of space.

The difference between the two views is readily illustrated. Consider a spacecraft flying by a planet. Newtonian physics says that the spacecraft's path is bent from a straight line by the gravitational force of the planet. According to Einstein, the spacecraft experiences no force and proceeds along what it "views" as a straight path; but because space is warped by the planet's mass, what we observe is a curved path. Einstein said the spacecraft feels no force; it just continues along in uniform motion obeying Newton's first law of inertia—but in a curved space-time. You can relate this to your own experience. Imagine driving down a highway using cruise control and your car comes to a curve. If the highway is properly banked, the car will automatically go around the curve without your touching the steering wheel and without the car changing its speed. The car continues along in uniform motion. Many children's toys and amusement park rides show this also. Modern rides, like roller coasters, sweep the customer along a curved track through space-time. Ride design can keep the force on the passenger's seat one g the whole time—no worry about falling out or spilling your Coke. Sometimes the rides have falls, bumps, and wiggles added in order to make them scary.

Gravitational attraction, said Einstein, is the result of curvature in space, not a mysterious force between objects. And, said Einstein, just as mass may be deflected by curvature of space, so too is light. Replacing the gravitational field by the curvature of space would have pleased Newton, who had not been satisfied with his laws even though they turned out to be so powerful as to dominate science for the succeeding three centuries. He wrote in a letter to Bentley: "That one body may act upon another at a distance through a vacuum without the mediation of any thing else, by and through which their action and force may be conveyed from one to another, is to me so great an absurdity, that I believe no man, who has in philosophical matters a competent faculty of thinking, can ever fall into it."

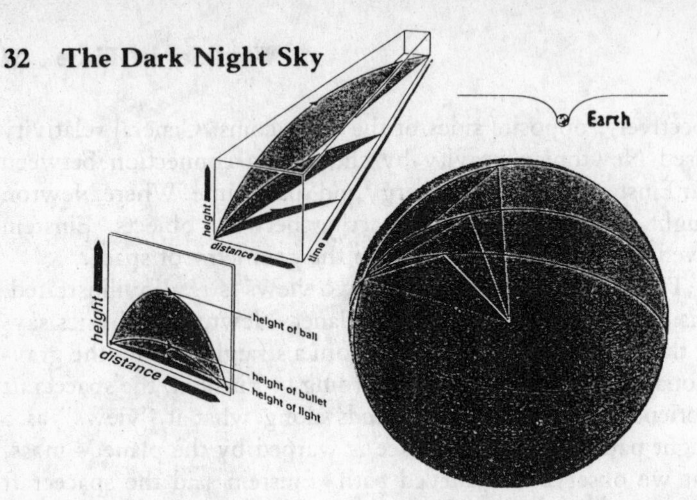

Gravity's rainbow. *Working from upper right:* The Earth's mass warps space-time in the same way as the Sun. A sphere with a light-year radius has about the same curvature as space-time at the surface of the Earth. All things continue their motion undisturbed—the principle of inertia. On the local surface of the space-time matching sphere, their paths are sections of great circles. Some things, like light or a speeding bullet, have great velocity, traveling a spatial distance in a short time; others, like a tossed ball, take a much longer time to travel the same spatial distance. Thus their circles have different angles in space-time and the arcs of their paths through space-time differ accordingly. When viewed in the laboratory, their paths look very different, making a gravity's rainbow of parabolic arcs. Viewed from the space-time perspective, their paths are simple and undisturbed identical arcs cut to different lengths. CHRISTOPHER SLYE

Physics took another giant step forward when Einstein merged Newton's laws of inertia and gravity through the concept of curved space-time. This achievement pointed toward our modern concept of the nonstatic universe. Imagine a universe with particles uniformly spread throughout it. Over time, in regions where space-time is positively curved, the particles will come together; where space-time is negatively curved, the particles will move apart, making voids. Only where space-time is flat will the particles maintain constant separation.

Very few nonscientists had heard of Einstein before his new theory was on the front page of every newspaper in the world. He was increasingly well known to physicists, who respected his contributions, particularly regarding the photoelectric effect, which he explained in terms of elementary quantum theory (and for which he would win the Nobel Prize in 1922). They were also intrigued by his relativity theories. Still, at age forty, his

The concept of curved space was proposed by mathematician Carl Friedrich Gauss (1777–1855) some two thousand years after Euclid wrote down the elements of flat-space geometry. Euclid synthesized what had been learned by Egyptian surveyors. The key to his geometry is the fifth postulate—two lines starting parallel remain equal distances apart or equivalently, the sum of the interior angles of a triangle is 180 degrees. Gauss arranged a survey of a triangle formed by the Harz mountain peaks Inselsberg, Brocken, and Hoher Hagen to see if the sum of the angles was 180 degrees. It was, to the surveyor's accuracy.

In 1826, Nikolai Ivanovich Lobachevski (1792–1856) developed an open, or equivalently negatively curved, geometry. The same geometry was independently developed by Gauss and János Bolyai (1802–60), an Austrian army officer.

On June 10, 1854, the great mathematician Georg Riemann (1826–66), at the age of twenty-eight, gave to the world the mathematical tools to define and calculate geometrical curvature. He also discovered positively curved-space geometry. He spent the rest of his life trying to use the idea of curved space to unify gravity, electricity, and magnetism. But he failed because he kept thinking about the connection between gravity, space, and curved space—not between gravity, space-time, and curved space-time.

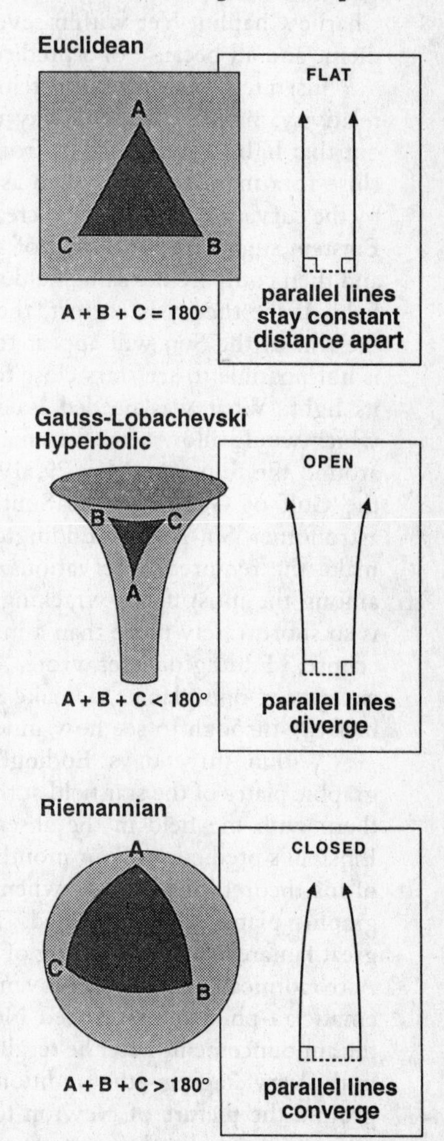

most brilliant years behind him, a scruffy but amiable eccentric who didn't wear socks, he was hardly a celebrity—hardly a man you would expect would go on to meet presidents, premiers, and Charlie Chaplin. Yet within several years he would meet all of them, and all because of a prediction that came true.

Just a few years after Einstein published his theory of general relativity, nature offered a way to test it. Einstein had pointed out that light traveling to us from stars should bend as it passes close to a massive body, such as the Sun. The light is diverted by the curvature in space-time created by the massive body. Look, Einstein suggested, at a field of stars in the absence of the Sun, and then compare the same field when the Sun blocks part of the field. If the theory is correct, then the position of stars close to the rim of the Sun will appear to shift. Normally, of course, it is not possible to see stars close to the Sun, as we are blinded by its light. What was needed, said Einstein, was a total eclipse, which would allow a brief moment's observation of the star field around the Sun. On May 29, 1919, such an eclipse was due in the Gulf of Guinea in the Southern Hemisphere. The British astronomer Sir Arthur Eddington organized an expedition to make the required observations. Total eclipse observations are among the most nerve-wracking in astronomy, because totality is so short: rarely more than a minute or two. "I did not see the eclipse," Eddington later wrote, "being too busy changing plates, except for one glance to make sure it had begun and another halfway through to see how much cloud there was."

Within three days Eddington had developed the photographic plates of the star field at the moment of eclipse, compared them with the field in the absence of the Sun, and confirmed Einstein's prediction. Four months passed before Einstein learned of his theoretical triumph, when detailed analysis of the photographic plates was completed. The results were announced to great fanfare at a joint meeting of the Royal Society and the Royal Astronomical Society, on November 6, 1919. The British mathematician-philosopher Alfred North Whitehead was present at the announcement. Later he recalled: "There was dramatic quality in the very staging—the traditional ceremonial, and in the background the picture of Newton to remind us that the greatest of scientific generalizations was now, after more than two centuries,

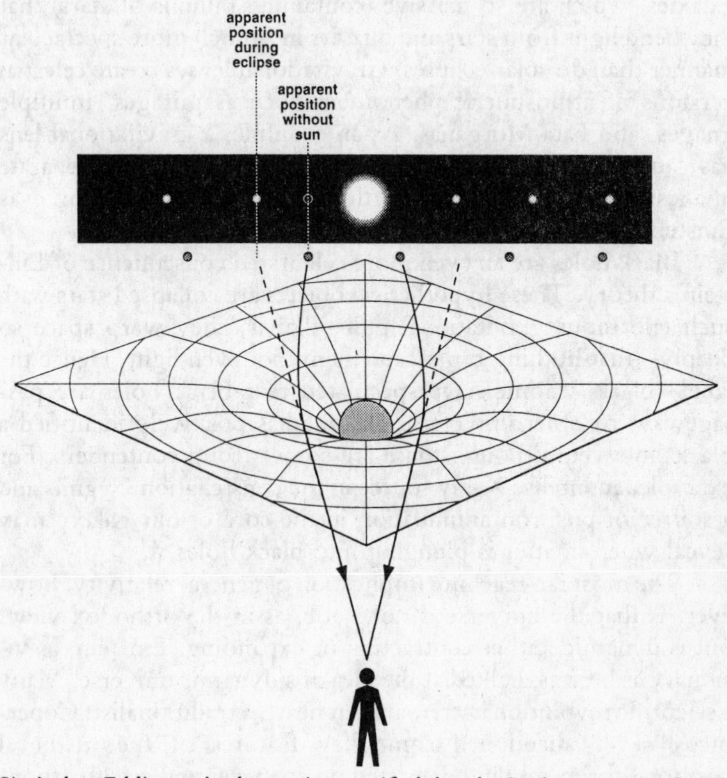

Sir Arthur Eddington's observations verified that the Sun's curvature of space "alters" the apparent position of a star, as Albert Einstein had predicted. CHRISTOPHER SLYE

to receive its first modification . . . a great adventure in thought had at length come safe to shore." Newspaper headlines ballyhooed the discovery: NEWTONIAN IDEAS OVERTHROWN, blazed one; SPACE "WARPED," declared another. The achievement would "overthrow the certainty of the ages," declared the London *Times*. "Epochmaking," said *The New York Times*. Einstein responded with aplomb. Had the experiment failed to support him, he would have been "sorry for the dear Lord—the theory *is* correct."

The consequences of Einstein's relativity theories are several and dramatic, not the least of which is the predicted existence of gravitational lenses. These are huge astronomical objects, such as galaxies, which are so massive (containing billions of stars) that they bend light from stars and quasars in a much more spectacular manner than do solar eclipses. Gravitational lenses create celestial versions of atmospheric phenomena such as mirages, multiple images, and Fata Morganas. As an example, a gravitational lens has caused light from a single quasar to split into five separate images. Because of its shape, this amazing sight is known as Einstein's cross.

Black holes are an even more celebrated consequence of Einstein's theory. These hypothetical objects are collapsed stars with such enormous gravitational pull—that is, they warp space so deeply—that nothing can escape them, not even light. Hence the label "black." Some have speculated that black holes are passageways to other universes. No one has positively identified a black hole yet, although there are some strong contenders. For example, an intense X-ray source in the constellation Cygnus and a source of positron annihilation in the core of our galaxy may reveal where matter is plunging into black holes.

The most far-reaching implication of general relativity, however, is that the universe is not static, as in the orthodox view, but is dynamic, either contracting or expanding. Einstein, as visionary as he was, balked at the idea of a dynamic universe. Many a scientific revolutionary remains, at heart, a traditionalist. Copernicus never abandoned some key features of the medieval cosmos—for example, he insisted on epicycles and circular (nonelliptical) orbits. Likewise, Einstein's thinking was so entrenched in the prevailing views that he rejected the radical notion of a dynamic universe.

One reason Einstein initially rejected this implication of his general relativity theory was that, if the universe is currently expanding, then long ago it must have started from a single point. All space and time would have been bound up in that "point," an infinitely dense, infinitely small "singularity." Hence it would be impossible to calculate what happened "before" the singularity, as any calculations would yield nonsensical results. The singularity would be an ultimate barrier to human knowledge, and this struck Einstein as absurd. He therefore tried to sidestep the

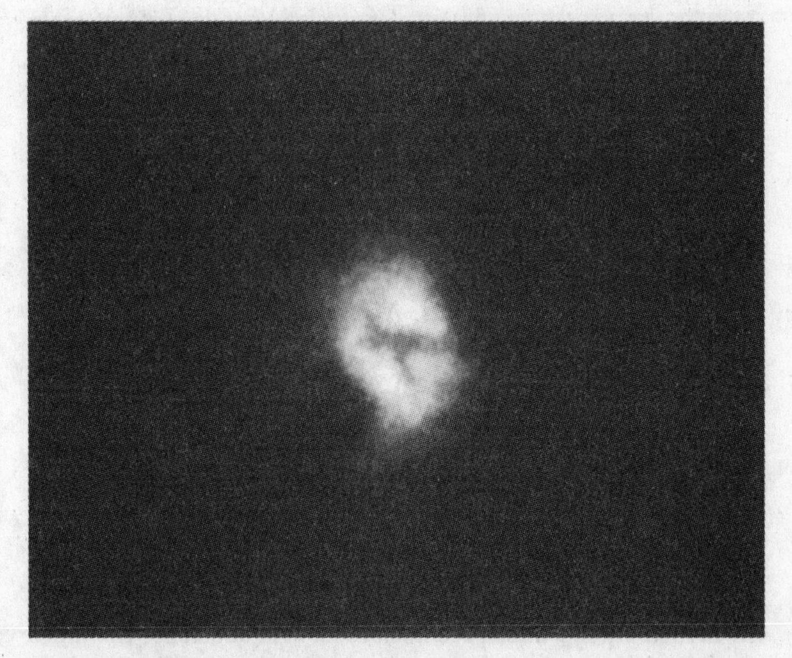

The silhouetted X structure may mark, within the nearby M 51 spiral galaxy, a potential black hole, with a mass equivalent to one million stars like the Sun. H. FORD (JHU/SPACE TELESCOPE SCIENCE INSTITUTE), THE FAINT OBJECT SPECTROGRAPH IDT; AND NASA

logic of his equations, and modified them by adding an arbitrary term known as a "cosmological constant." This term represented a force, of unknown nature, that would counteract the gravitational attraction of the mass in the universe. That is, the two forces would cancel each other, yielding a static cosmos, neither expanding nor shrinking. Einstein had no evidence for the existence of a cosmological constant; it is the kind of rabbit-out-of-a-hat idea that most scientists would label *ad hoc*. And in Einstein's case it represented his philosophy embodied within his mathematical equations.

Ironically, Einstein's approach contained a foolishly simple mistake: His universe would not be stable. The slightest decrease in distance between objects would cause their mutual gravitational pull to be greater than the cosmological constant repulsion, hence they would begin moving together (or apart if their distance increased slightly). His universe was inherently unstable, like a pencil balanced on its point.

There was a second philosophical reason that influenced Einstein to reject the equations' conclusion—one that had guided him in creating his equations. In creating his relativity theory Einstein had to reject Newton's concept of absolute space and time. Newton was the revolutionary arguing that empirical evidence overruled philosophical arguments. At the time, Newton's archrival, Gottfried Wilhelm von Leibniz (1646–1716), argued that there was no philosophical need for any conception of space apart from the relations between material objects. Newton carried the scientific world, but philosophers continued to explore the possibilities Leibniz suggested, with Immanuel Kant (1724–1804), proponent of the island universe and absolute space, and Leonhard Euler (1707–83) opposing Bishop George Berkeley (1685–1753).

In 1721 Berkeley published *Motion,* a reversion to the Aristotelian belief that space exists because of the matter in it. Aristotle had argued that space is clothed with substance that both gives it reality and moderates the motion of objects. Space by itself could not exist, so there was no vacuum (hence the Greek atomists were wrong to say there is nothing except "atoms and the void"). Berkeley amplified Aristotle's argument that space by itself is emptiness and therefore nothing, saying: A single body in an otherwise empty universe has no measurable motion of any kind. Two bodies can define relative motion on the axis formed by the line between them. Four bodies can define motion in three dimensions, but it would take more than four bodies to define rotation. In this way, the properties of what Newtonians called absolute space were the result of the material content of the universe.

Austrian philosopher Ernst Mach (1838–1916) expressed similar ideas to Bishop Berkeley's. He went a little further to say that adding a few tiny dust specks to space could give reference points for motion, but filling the universe with material would create a more substantial framework. A uniform distribution of matter should create a uniform space. In 1893 Mach stated the hypothesis: The influence of all the mass in the universe deter-

Time exposure of the night sky showing circular star tracks due to the Earth's rotation. DAVID NUNUT/SKY AND TELESCOPE

mines what is natural motion and how hard it is to change. Does this remind you that natural motion is inertia? It should.

We can examine Mach's hypothesis for ourselves. If we look at the stars and their cousins in the distant galaxies, they appear to rotate overhead—once every twenty-four hours. However, compared to the plane of the Solar System, the observed rotation of the galaxies is less than one second of arc per century. If we go to the Earth's poles and hang a pendulum on a long cord and let it swing freely, then the plane of its swinging stays fixed relative to the distant galaxies rather than rotating with the Earth. In other words, its course of natural motion is aligned with the distant stars. Given this observation, either (1) Newtonian absolute space-time defines natural reference for motion, and while the Earth rotates and orbits in it, the galaxies happen to be at rest in it, or (2) Mach's hypothesis is right and the average of all matter defines our inertial reference frame.

Einstein was deeply influenced by Mach, with whom he corresponded. Einstein labeled the hypothesis "Mach's principle" and used it as a guide when developing general relativity. In 1916

Karl Schwarzschild found the first solution to Einstein's general relativistic equations—specifically, a solution for a single static mass in an otherwise empty flat space. It is the solution now used for calculation of the Sun's bending of light, shown schematically in the illustration on page 35, and other effects in the Solar System. Einstein was not impressed and was slightly distressed by Schwarzschild's solution—holding to Mach's principle, Einstein thought that a mass by itself in space had no meaning. The Schwarzschild solution gave the right answer for the Solar System, but only because—Einstein argued—the rest of the matter in the universe provided the flat space-time background.

This motivated Einstein to construct the first general relativistic cosmological model in 1917. His goal: a static universe that obeyed Mach's principle. His general relativity equations would not allow a static universe, so he added the cosmological constant. In his model, space had the geometry of a three-dimensional sphere—finite but with no border. Galaxies in this sphere kept fixed spatial locations and just traveled along in time. Einstein was convinced that his solution was the only possible model without the dreaded singularity. However, in 1919 Willem de Sitter of Holland came up with a solution, including the cosmological constant, in which a universe without matter would *expand.*

Einstein detested de Sitter's model because it contained no matter. How could space make sense without matter? Matter *defined* space. Einstein had thought general relativity contained Mach's principle. Einstein and the world would eventually learn that Mach's principle is not built into the general relativity equations but, rather, requires additional conditions to be valid.

For the next decade cosmological controversy centered on which of the two models, Einstein's or de Sitter's, was right. In 1922, the Russian Alexander Friedmann showed that Einstein's equations have a class of solutions representing a uniform distribution of expanding matter. Friedmann, who had the clearer insight that expansion is a key aspect of a relativistic universe, died in 1925, before seeing his vision fully accepted.

In 1932, Einstein and de Sitter collaborated to find a specific Friedmann solution—one in which space is flat (with zero cosmological constant) but expanding, known as Einstein–de Sitter space. This solution appears to be a very accurate representation of what we can see of our universe.

When Einstein described his revolutionary cosmological model before a gathering of the Berlin Academy of Sciences in 1917, he explained his rationale for adding the cosmological constant to the equations: "That term is necessary only for the purpose of making possible a quasi-static distribution of matter, as required by the fact of the small velocities of the stars." The mathematical device expunged the implications of an expanding universe from the equations of general relativity, and left intact the old notion of a static universe. There was no strong observational evidence of an expanding universe at the time, and there was philosophical comfort in believing in a static universe—for example, one didn't have to address the question of what happened "in the beginning." So Einstein, despite his genius, clung to the static universe although his equations predicted a dynamic one, as empirical evidence would demonstrate. Uncharacteristically, Einstein failed to trust the logic of his equations, something he later described as "the biggest blunder of my life."

Chapter 3

The Expanding Universe

Edwin Powell Hubble is one of the great heroes in the scientific odyssey that led to the modern big bang theory. Born in 1889 in Marshfield, Missouri, Hubble became a Rhodes scholar at Oxford University and a master athlete. He fought in the First World War and taught high school (and was so beloved by his students that they dedicated an annual yearbook to him). He also studied for the law and practiced briefly in Kentucky, but quickly lost interest in it. His passion was astronomy. Despite his unorthodox academic background, he managed to gain access to the (then) biggest telescope in the world, the Mount Wilson Observatory in southern California. Before the large-scale electric lighting and the suburban sprawl of Los Angeles County obscured its view, Mount Wilson, with its hundred-inch telescope, enjoyed clear, dark skies and had an unparalleled window onto the heavens.

Hubble's extraordinary dedication and work made him one of the most celebrated astronomers of the day, and in 1948 he made the cover of *Time* magazine, on which his visage looked appropriately somber while, in the background, a giant finger pointed toward the stars.

Hubble has been described as an "extraordinarily exact and careful scientist who normally refrained from assertions that were not well supported by evidence." Yet his science was filled with passion: He spent hundreds of bone-numbing hours in the ob-

Edwin Hubble, pointing to a photograph of the Andromeda Galaxy. THE HUNTINGTON LIBRARY

server's cage at the Mount Wilson telescope. Anyone who has spent time in "the cage" knows what an extraordinary effort of will it can demand: total concentration, and an ability to suppress shivers in the constant chill, lest you vibrate the telescope. Hubble subjected himself so slavishly to this discomfort because he wanted to transform the science of astronomy. In his graduate thesis he had called on astronomers to investigate whether the spiral nebulae were located within our own galaxy or outside it, galaxies in their own right. Conventional wisdom held that the nebulae were objects within our galaxy; indeed, that the galaxy constituted the virtual extent of the universe. If, as Hubble suspected, the nebulae were extragalactic, discoveries of mythic proportions would be possible.

I know how Hubble felt. In the early 1970s, I chose to work on measuring cosmic background radiation partly because I knew this: Whatever we learned would be *fundamental*. Regardless of what we found, our observations would tell us about the early universe.

Night after night Hubble photographed the nebulae, devoting himself to his goal so completely that he was perceived (probably with justification) as being arrogant and elitist. He pored over the images on photographic plates and strained to perceive stars within the nebulae, and occasionally managed to do so, depending "on what I had for breakfast." He benefited especially from the aid of Milton Humason, a onetime mule driver who became an observatory assistant and took many of the most important images of the nebulae.

In 1924 Hubble spotted something important: It was a very special kind of star in the Andromeda Nebula, a vaguely oblong glow near the constellation Cassiopeia (the one shaped like a crooked *W*). The nebula is easily seen with binoculars. If you live, as I do, near latitude 40 degrees N, then the nebula is directly overhead at midnight in mid-January. It's quite large—about six degrees across, or about ten times the size of the Moon. The special star Hubble had seen in Andromeda was a Cepheid variable; such stars are unusual because their luminosity oscillates with a regular period. On his photograph of the Cepheid, he excitedly wrote, "VAR!" The presence of the star would allow Hubble to do what astronomers had struggled to do for years: reliably measure the distance to a nebula. The measurement would reveal whether the spiral nebulae were truly extragalactic or were merely wayfarers within our own galaxy.

Measurement of astronomical distances had long frustrated astronomers. The only trustworthy technique involved the phenomenon of parallax. Here's a simple example of parallax: Hold your finger up to your eyes and shut one eye, then open it and shut the other eye. Notice how your finger seems to move back and forth in relation to the background (the wall, for instance); the magnitude of the finger's shift is its parallax. The nearer the finger is to the eye, the bigger the shift, or parallax; the farther from the eye, the smaller the parallax. The parallax effect is caused by seeing the finger from the different positions of the two eyes. You can calculate the distance of your finger from your eyes from the simple geometry of the triangle formed by the two eyes and the finger. Astronomers can likewise determine the distance of heavenly objects. An object is observed from two different places (two widely separated observatories), and its parallax against the background of stars is determined. The triangle here is formed by the two observatories and the heavenly body. Again, simple geometry gives the distance.

For nearby objects—planets within the Solar System—this is an excellent way to measure their distance. But for far-flung stars, the distance between two Earth-based observatories is insufficient to produce a measurable parallax. For nearby stars, astronomers exploit the Earth's movement around the Sun. Make an observation of the target star in, say, January, then make a second observation six months later, when the Earth has moved on its orbit to the opposite side of the Sun. The distance between the two observations is 187.5 million miles, the diameter of Earth's orbit around the Sun. The triangle now is formed from the two sides of Earth's orbit and the star—and simple geometry works again. Unfortunately, the parallax technique requires extremely precise observations and works only for the nearest stars. For distant stars, 187.5 million miles is too short a distance to provide the base of the triangle.

In the early 1910's, Henrietta Leavitt, a deaf woman working at Harvard College Observatory, discovered a radically new way to measure cosmic distances. She realized that the absolute brightness of Cepheid variable stars is related to the time period within which they brighten and dim: The brighter the star, the longer its cycle. (Absolute brightness is the true brightness of a star, as it would appear to a nearby astronaut. Apparent brightness is how bright it appears to us from however many light-years away it is.) Using a simple calculation based on the assumption that the intensity of light diminishes with the square of the distance, one can calculate how distant the star must be in order to explain its apparent brightness.

Hubble exploited Leavitt's technique to measure the distance to Andromeda, based on the periods and apparent brightness of its Cepheid population. The distance, he concluded, was 800,000 light-years, which is ten times the typical distance to stars within our galaxy.* Andromeda was clearly beyond the realm of the

*This estimate has since been revised upward, to about 2 million light-years. That means light left the Andromeda Galaxy 2 million years ago, about the same time the genus *Homo*, to which humans belong, began evolving in Africa.

Milky Way and, hence, must be a distant galaxy in its own right. Hubble had achieved his goal: Astronomers could now be certain the nebulae are, indeed, separate galaxies. And there evidently were countless such galaxies beyond our own, meaning the cosmos must be far bigger than suspected earlier. With Hubble's discovery, a new age in astronomy had dawned.

The discovery that the universe is vast—effectively infinite—was revolutionary enough. Cosmology's concept of a cozy universe, not much greater in extent than our own visible galaxy, was banished to the history books. But Hubble went further. Building on his discovery by exploiting a technique developed earlier by the American Vesto Melvin Slipher, Hubble then struck at the centuries-old belief that the universe is static—the notion to which Einstein clung so tenaciously.

In the 1910's, Slipher was hard at work in the American desert, using Lowell Observatory to monitor nebulae. The observatory sat atop Mars Hill in the town of Flagstaff, Arizona, within walking distance of cowpokes and saloons. Here, on the eve of the First World War, Slipher uncovered the first direct evidence for the expanding universe, but didn't realize it.

A native of Mulberry, Indiana, Slipher was twenty-five when he came to the observatory in 1901. He had been hired by its founder, Percival Lowell, one of the most controversial figures in the history of astronomy. Lowell was the scion of a wealthy Boston family that included famed industrialists, academics, politicians, and literati. He had attended Harvard, studied mathematics, and served as a U.S. diplomat in the Far East. He is best known for his interpretation of the enigmatic lines that pattern the surface of Mars. Lowell thrilled the *fin de siècle* world with his claim that the lines were canals, constructed by Martians to pipe water from the polar ice caps across their desert world. Great canals were being constructed across Suez and Panama at this time, and were regarded as great engineering feats of high civilization. Imagine the technological prowess of a civilization able to construct canals that girdled an entire planet!

Like the eighteenth-century French astronomer Charles Messier and the nineteenth-century English astronomer William Herschel, Lowell was also interested in the nature of nebulae, which he suspected were clouds of dust and gas that, over time, had condensed into new (and perhaps habitable) planetary systems. To learn whether this was so, in 1909 he instructed Slipher to make spectral observations of nebulae.

Slipher started by photographing the spectrum of the Andromeda Nebula, which yields something of a visual "signature" for the galaxy. White light, when passed through a prism, splits into its component colors, the classic rainbow spectrum of red, orange, yellow, green, blue, indigo, and violet. Lines in the spectrum reveal the chemical composition of stars or galaxies. Late in 1912, after an extensive series of photographs climaxing on New Year's Eve, Slipher examined four spectrograms of Andromeda and realized their spectral lines were in the wrong place: The lines were shifted toward the blue end of the spectrum. The apparent cause was a phenomenon known as a Doppler shift, a phenomenon named after the Austrian physicist Christian J. Doppler (1803–53).

The Doppler effect is most readily explained by considering what happens to the sound of a train's whistle as it passes by. During its approach, the sound is relatively higher pitched, and when it passes and recedes into the distance it becomes lower pitched. The whistle on a stationary train produces sound with wave crests that pass you at a constant rate. But when the train rushes toward you, every new wave crest is created closer to you than if the train were stationary. The crests of its sound waves are packed together by the motion of the train toward you, which effectively increases the frequency (a higher pitch) of the sound. As the train recedes down the track, the situation is reversed: The crests are produced at a greater and greater distance and the sound waves reach the ear at a slower rate—its frequency is decreased and the train's whistle pitch drops.

The same effect occurs with light emitted by a moving object. Color is the optical counterpart of pitch: In the visible region of the electromagnetic spectrum, the color red is a lower frequency (like the low-pitched whistle), while blue is a higher frequency (the high-pitched whistle). Hence, a star approaching us should appear bluer—that is, its frequency should shift toward the blue end of the spectrum—and a star moving away from us should shift toward the red end. The former is said to be blue-

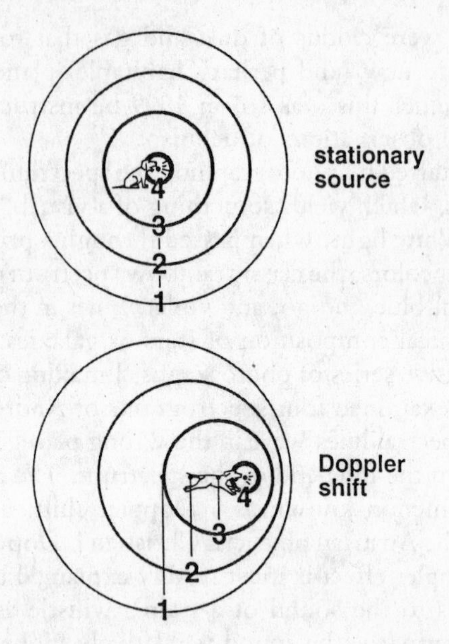

The Doppler Dog. In the Doppler effect, a moving source changes its pitch relative to a stationary one. The Doppler Dog, who barks at a constant rate, demonstrates the Doppler effect. Each time he barks, a wavefront burst of sound goes out spherically from his mouth. Everyone around him hears the barks arriving equally spaced. If the dog runs while he barks, his mouth moves and the center of each expanding sphere of sound is displaced. Listeners in the direction the dog is moving hear the barks with shorter spaces between them, so they are higher in pitch. Listeners in the direction he is leaving hear a lower pitch, because there is more time between barks. CHRISTOPHER SLYE

shifted, the latter redshifted.* Nineteenth-century astronomers began analyzing Doppler shifts in the spectra of stars. That's how they learned that some stars are moving across space at tens of miles per second, which seemed pretty fast back then!

*The frequency or wavelength shift is determined by measuring shifts in thin, dark lines on the spectrum. These "absorption" lines are called Fraunhofer lines after Joseph von Fraunhofer (1787–1826) of Germany. The lines are caused by intervening cooler gas (such as the star's atmosphere) that absorb certain frequencies of light. Dutch physicist Pieter Zeeman (1865–1943) analyzed "emission" lines, which are thin, *bright* lines produced by the same gas if it is hot. Nineteenth-century scientists realized that different elements produce different arrangements of lines, a fact later explained by quantum physics.

Hubble's velocity vs. distance plot of 1929.

The black discs and full line represent the Hubble law fitted to the nebulae individually. The circles and broken line represent the solution combining the nebulae into groups.

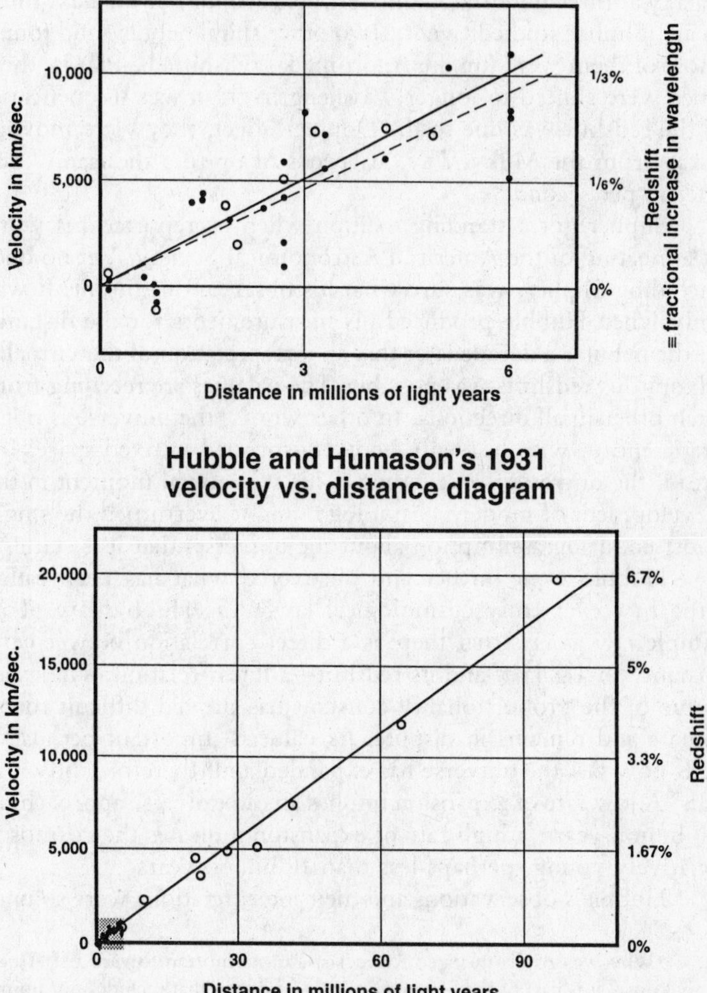

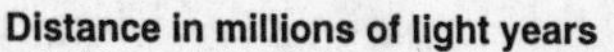

Hubble law data. The upper plot shows Hubble's original data on distance to nebulae (galaxies) and their red shift (or velocity using the Doppler effect relation). The lower plot shows Hubble's original data in the shaded region and the new data collected with Milton Humason. The new data provide a convincing argument for the linear relationship between distance and red shift. CHRISTOPHER SLYE

Early in 1913, Slipher told Lowell that Andromeda displayed a pronounced blueshift: The nebula was speeding toward our galaxy at three hundred kilometers per second. For the next three years, Slipher studied twenty-two other spiral nebulae and found most of them were (unlike Andromeda) redshifted—that is, their lines were shifted to longer wavelengths (to lower frequencies). If the redshift was due to the Doppler effect, they were moving away from the Milky Way, at speeds of up to a thousand kilometers per second.*

Slipher got a standing ovation when he reported his work at a meeting of the American Astronomical Society. Yet no one, including Slipher, was sure what his observations meant. It was only when Hubble produced his measurements for the distance of the nebulae a decade later that an interpretation of the virtually ubiquitous redshifts was possible: The galaxies are receding from each other in all directions. In other words, the universe is not a static entity, with heavenly bodies suspended in fixed space. Instead, the universe is expanding. This was a vital moment in the development of modern cosmology, and it overturned the single most enduring assumption about the universe: that it is static.

Hubble went further and uncovered what has been called "the first ever truly cosmological law": the Hubble law. This simple law asserts that there is a direct correlation between the distance to a galaxy and its redshift—a linear relation. The exact value of the proportionality constant has proved difficult to estimate and remains in dispute. Its value is important because it tells how fast the universe has expanded and, therefore, how old it is. A low rate of expansion implies an old universe approaching 20 billion years; a high rate of expansion indicates the cosmos is relatively young, perhaps less than 10 billion years.

Hubble's observations and their interpretations were so une-

*Why was Andromeda so iconoclastic in its motion toward us? As we now know, it is part of the "Local Group" of galaxies that includes our home, the Milky Way. The Local Group includes a few dozen fair-sized galaxies that are united by their mutual gravity and, therefore, aren't dispersed by cosmic expansion. Eventually Andromeda and the Milky Way may collide; astronomers have seen many other examples of such collisions in space. Fortunately, stars are widely dispersed in galaxies, hence one galaxy can pass through another with minimal wear and tear—like one cloud passing through another.

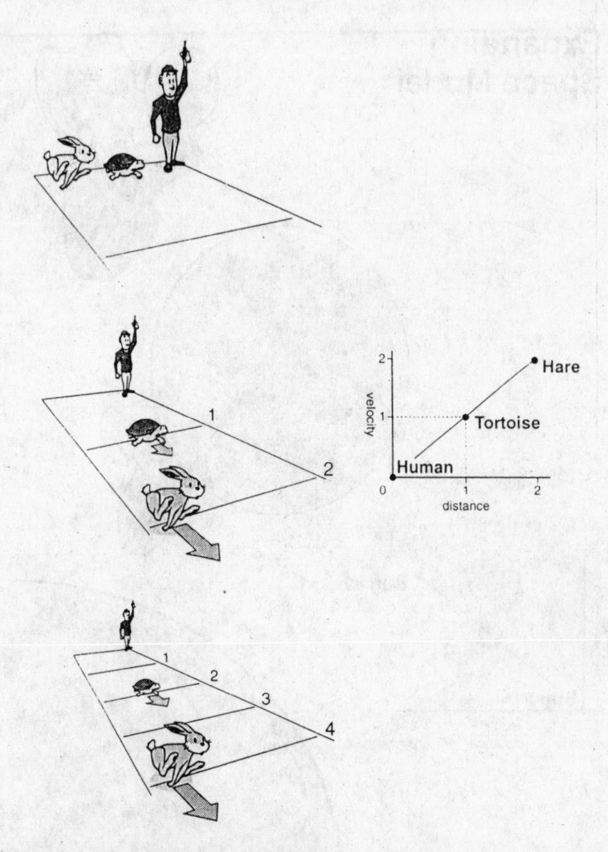

Race model/giant explosion model. The diagram shows how the Hubble law of linear relationship between velocity and distance occurs in a race or a giant explosion. The racer (or debris) moving the fastest goes the farthest. The distance traveled is just the speed multiplied by time elapsed since the start. This automatically gives the Hubble law but is a misleading picture of how we now view the big bang. CHRISTOPHER SLYE

quivocal that they quickly convinced even the most ardent adherents to the static universe idea. The observational evidence was so powerful it was impossible to ignore. Einstein was among the converts, and described as his "greatest blunder" his prior inability to realize that his equations implied an expanding universe. In 1930, Einstein and his wife, Elsa, visited Mount Wilson, where Hubble gave them a tour of the observatory. Elsa was told that the hundred-inch telescope was used to study the structure

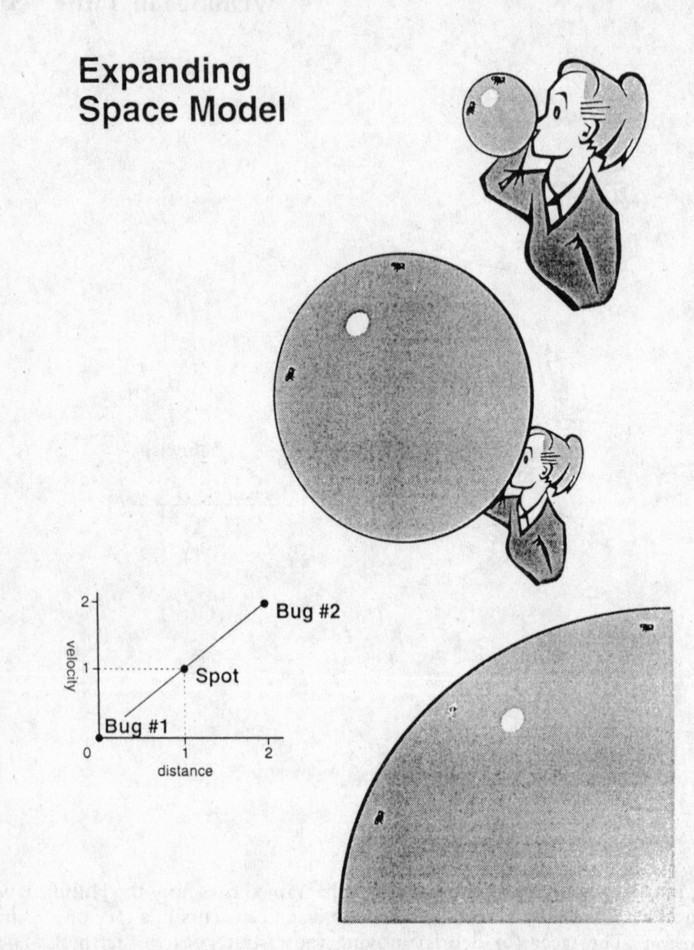

The distance between the bugs and the spot expands with the balloon, just as the distance between galaxies increases during the expansion of space. This does not necesssarily mean galaxies move *through* space; as space expands, it carries the galaxies with it. CHRISTOPHER SLYE

of the cosmos. "Well, well," she replied, "my husband does that on the back of an old envelope."

Eventually it was the combination of the work of these two men that led to our present understanding of the universe. The crucial question remained: What had *caused* the expansion?

It often happens in science that, through the confluence of circumstance, the time is ripe for an important conceptual insight.

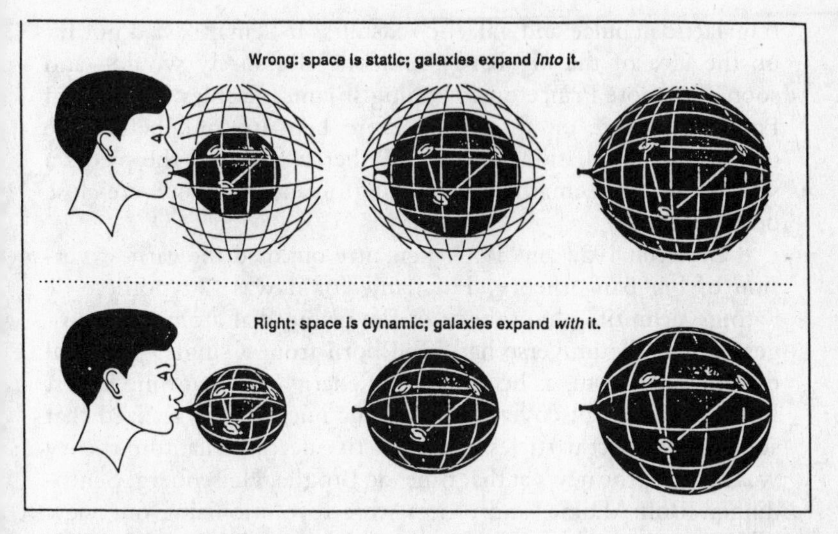

Galaxies expanding into space (*top*) versus galaxies expanding *with* space, not into it. SKY AND TELESCOPE

So it was for the priest and astronomer Georges-Henri Lemaître, who, at the turn of the second and third decades of this century, conceived the notion that ultimately became the big bang theory.

Born on July 17, 1894, at Charleroi, Belgium, Lemaître served in the First World War, witnessed, among other horrors, a chlorine gas attack, and won medals for valor. An aggressive personality, he was expelled from a military training class for challenging the teacher's incorrect answer to a question in ballistics. During lulls in the fighting, Lemaître read a book on cosmology by the great theorist Jules-Henri Poincaré and became entranced with its challenges. In the early 1920's Lemaître enrolled at Cambridge University and subsequently at Harvard University to study astronomy.

It was an auspicious time. Einstein had recently produced his theory of general relativity and was struggling to avoid its implication of a nonstatic (probably expanding) universe; Alexander Friedmann and Willem de Sitter were interpreting Einstein's equations and accepted their implication of a nonstatic universe; Slipher was accumulating data about galactic redshifts; and Hubble was about to make history by confirming the reality of extragalactic nebulae and galactic redshifts. If Lemaître had not hit on the idea of the big bang, someone else surely would—and soon. To quote Princeton cosmologist James Peebles: "Weyl and Friedmann were on the track before Lemaître and before the observational situation was ripe; Robertson had all the pieces a year or so after Lemaître, and Eddington and Tolman were close behind him."

Between 1927 and 1933, Lemaître outlined the earliest version of big bang theory. His name for it was "Hypothèse de l'atome primitif" (hypothesis of the primordial atom). He suggested that the universe had been born from a single, primeval quantum (or atom, as he called it) of energy. By that time Ernest Rutherford had discovered the atomic nucleus and realized that some nuclei eject particles by radioactive decay. Quantum theory was much in the news at that time: de Broglie, Heisenberg, Schrödinger, Bohr, Dirac, and others were revolutionizing our view of the microworld, and the foment surely helped inspire Lemaître.

In Lemaître's view, the primordial atom had begun dividing, over and over like bacteria in a dish. Eventually it spawned all the matter of the present universe. Space and time unfolded as the nuclei proliferated. Quantum self-replication took place at an explosive rate. "The evolution of the universe can be compared to a display of fireworks that has just ended: some few wisps, ashes and smoke," he wrote in the late 1920's. "Standing on a well-chilled cinder, we see the slow fading of the suns, and we try to recall the vanished brilliance of the origin of the worlds."

Lemaître unsuccessfully tried to interest Einstein and de Sitter in the primordial atom. He sought out Einstein at the Fifth Solvay Conference in Brussels in 1927 to plead his case. Einstein was uncharacteristically brusque and snapped: "Your calculations are correct, but your physical insight is abominable." Lemaître's old teacher Arthur Eddington was repelled by talk of cosmic beginnings: "It has seemed to me that the most satisfactory theory would be one which made the beginning *not too unaesthetically abrupt*," Eddington harrumphed (his italics).

Lemaître described his ideas in the May 9, 1931, issue of *Nature*, in a letter that one scholar has called "the charter of the Big Bang theory." Lemaître began by recalling that Eddington had scorned talk of cosmic origins because, philosophically,

> the notion of the beginning of the present order of Nature is repugnant to him. I would rather be inclined to think that the present state of quantum theory suggests a beginning of the world very different from the present order of Nature. Thermodynamical principles from the point of view of quantum theory may be stated as follows: (1) Energy of constant total amount is distributed in discrete quanta. (2) The number of distinct quanta is ever increasing. If we go back in the course of time we must find fewer and fewer quanta, until we find all the energy of the universe packed in a few or even in a unique quantum.
>
> Now, in atomic processes, the notions of space and time are no more than statistical notions; they fade out when applied to individual phenomena involving but a small number of quanta. If the world has begun with a single quantum, the motions of space and time would altogether fail to have any meaning at the beginning; they would only begin to have a sensible meaning when the original quantum had been divided into a sufficient number of quanta. If this suggestion is correct, the beginning of the world happened a little before the beginning of space and time. I think that such a beginning of the world is far enough from the present order of Nature to be not at all repugnant.

Lemaître's letter resulted in a story in the May 19, 1931, *New York Times*, headlined: LEMAÎRE SUGGESTS ONE, SINGLE, GREAT ATOM, EMBRACING ALL ENERGY, STARTED THE UNIVERSE.

While cosmologists debated cosmic beginnings, the world was suffering through the Great Depression. News media tried to cheer readers' spirits with exciting (sometimes exaggerated) reports on groundbreaking discoveries in science, especially astronomy. As a result, Lemaître became a minor media celebrity; although he wasn't as celebrated as Einstein, newspapers ran photos of them together. By this time Einstein realized he had dismissed too quickly the young priest's idea and began referring to it as the "most pleasant, beautiful and satisfying interpretation" of astronomical phenomena. The *Times* reassured readers (who no doubt recalled the Scopes "monkey trial" several years earlier) that Lemaître perceived "no conflict between science and religion" and was "one of the best mathematical physicists alive. . . . Just now his expanding universe is so popular that Einstein's static, cylindrical model looks as old-fashioned as a high-wheel bicycle."

Key to the acceptance of the primordial atom hypothesis had been Eddington, Lemaître's mentor and the most powerful figure

The early 1930's: Albert Einstein, whose special and general theories of relativity implied a radically new cosmology, and Georges-Henri Lemaître, the Belgian cleric and astrophysicist who pioneered the idea of the big bang. BROWN BROTHERS

in cosmology at the time. In 1927 Lemaître had given Eddington a manuscript outlining the hypothesis, but Eddington was so cool to the idea that he filed it away without reading it carefully. When Eddington recanted, five years later, he retrieved the manuscript and had it translated from French into English and published in the proceedings of the Royal Astronomical Society. The move represented an essential stamp of approval. The big bang theory (not yet so named) had arrived.

Progress in cosmology—as in all sciences—is nurtured by a constant interplay between theory and experiment (or observation). Lemaître's theory proved to be popular, but how could it be proven—that is, tested—scientifically? The products of the big bang are all around us in the universe we see today. But that universe is vastly complicated; scouring it for signs of a primordial event that happened some 15 billion years after it occurred is likely to be futile. A time traveler would see the universe become progressively simpler as he approached the big bang, and it is among these simpler manifestations of the event that clues must be sought. Lemaître realized this, and suggested that high-energy cosmic rays were relics of the big bang, not unlike sparks from an explosion. He was wrong, however, as these particles are almost certainly generated by processes in our galaxy. Confirmation of the big bang would have to wait for another thirty-three years.

Nonetheless, by the early 1930's, the idea and theory of expanding space had been finally and firmly introduced into cosmology. Unfortunately, many nonscientists believed—and continue to believe—that the big bang was an explosion that spewed matter into empty space. The name big bang does not help. The early interpretation of the cosmological redshift as a Doppler shift implanted in popular consciousness the idea that the galaxies were moving out into a preexisting space. This explosion/motion picture has led to many confusing and conflicting conundrums. Einstein introduced a clearer vision, namely, space and time are not the absolutes set by Newtonian physics but, rather, have variable properties of their own. In particular, space-time has curvature and, as Einstein reluctantly had to admit, changes in scale with time. We live in a universe whose space is currently expanding. The cosmological redshift is due to the stretching of light by the expansion of space. The light from distant galaxies takes longer to reach us and thus is stretched to longer wavelengths by expanding space than is light from nearby galaxies. Hence, expansion of space produces the Hubble law. Consider an idealized universe in which the galaxies are not moving relative to the space and other material near them. The expansion of space makes the distance between galaxies grow in proportion to the distance between them. If space expands at a constant rate, this relationship is the linear Hubble law—the redshift is proportional to distance. At a sufficiently large separation distance, two galaxies, neither of which is moving relative to its locale, have a separation distance increasing more quickly than light can travel between them. Without understanding that space is expanding, one would wrongly think that the galaxies were moving apart faster than the speed of light even though neither was moving.

The advent of big bang theory not only offered a dramatic new perspective on the origin of the universe, but it provided a possible resolution to one of cosmology's most persistent questions—where did the elements come from? How had nature fashioned the ninety-plus natural occupants (hydrogen, helium, carbon, oxygen, iron, and so on) of Mendeleyev's table of elements? Much of the world with which we are familiar—the rocks, animals, and plants around us—is composed of elements heavier than hydrogen and helium. The universe as a whole, however, is dominated by two elements: helium, which represents almost 25 percent of all matter, and hydrogen, close to 75 percent. All other elements total just 1 percent. How did this distribution of elements come to be? Lemaître's big bang theory would provide an answer—or at least a partial one.

Until the turn of the twentieth century, the question had seemed unanswerable, not least because no one knew the internal structure of atoms. In 1896 Henri Becquerel (1852–1908) discovered radioactivity. Ernest Rutherford (1871–1937), a New Zealander transplanted to England, arrived at Cambridge University in 1897 just in time to witness J. J. Thomson discover the electron. Rutherford began the study of radioactivity and won the Nobel Prize in 1908 for his work. Using a radioactive source to bombard a thin sheet of gold foil, Rutherford discovered the nucleus in 1910, and so pioneered subatomic physics. He found that the atom was almost entirely empty space. It contains a tiny, positively charged nucleus composed of protons, orbited by even tinier, negatively charged electrons. (The other constituent of the nucleus, the neutron, was discovered two decades later, in 1932, by Chadwick.)

An unusual fact was noticed as early as 1917 by the American chemist William Draper Harkins. Excepting hydrogen, he found, elements with even atomic numbers* (2, 4, 6, etc.) were far more abundant than those with odd atomic numbers (3, 5, 7, etc.). Was this a clue to how atoms formed? Harkins speculated that elements had been generated by the merging—fusion—of their atomic nuclei, and that even-numbered elements were more abundant either because they formed more easily or were more stable.

The notion of fusion as an important cosmological process was strengthened when, in 1925, the Harvard astronomer and department chair Cecilia Payne-Gaposchkin (1900–1979) showed that the Sun consists almost entirely of the lightest element, hydrogen. Theorists concluded that the Sun generates its heat not via atomic fission as had been supposed, but by fusion—specifically, the merging of two hydrogen nuclei (which have one proton apiece) into a helium nucleus (two protons). Intense heat is needed to initiate the fusion process because the two protons, both bearing positive charges, repel each other. Once under way, however, nuclear fusion releases immense energy, which is why physicists have been attempting for decades to harness nuclear fusion for energy generation here on Earth, so far unsuccessfully on any useful scale. In the 1930's, the German physicist Hans A. Bethe described the theory of nuclear fusion as the source of the Sun's heat.

Ideas from scattered disciplines were coming together, like separate roads meeting at an intersection. If nuclear fusion could explain stellar heat, then might it also explain the origin of elements? Might stellar heat, or some other heat source, have fused protons together in different aggregations, forming all the elements in the cosmos? For example, by merging helium nuclei,

*Atomic number is related to the number of protons in an element; hydrogen is the lightest element, having just one proton. Helium has two protons, lithium three, and so on.

with two protons apiece, one could get heavier elements containing protons in multiples of two: 4 (beryllium), 6 (carbon), 8 (oxygen), 10 (neon), and so on. That might explain why even-numbered elements are so common. "The [particles] constituting [helium's] nucleus must have been assembled at some time and place; and why not in the stars?" Eddington wrote in his 1927 book *Stars and Atoms*. "I am aware that many critics consider the conditions in the stars not sufficiently extreme to bring about the transmutation—the stars are not hot enough. The critics lay themselves open to an obvious retort; we tell them to go and find *a hotter place*."

But by the 1930's, a "hotter place" was available, at least on paper: Lemaître's primeval atom.

In 1938, after concluding that the interior of the Sun and other stars wasn't hot enough to fuse light elements into an abundance of heavier elements, the researcher Carl Friedrich von Weizsäcker suggested that a superhot primordial "fireball" might have done so instead. As he wrote in a European scientific journal:

> One may therefore presuppose a great primeval aggregation of matter perhaps consisting of pure hydrogen. As it collapsed into itself under the influence of gravity and thereby raised its central temperature, it came finally into a state in which nuclear reactions took place in its interior. . . .
>
> How large should one imagine the first aggregation to have been? Theory sets no upper limit, and our fancy has the freedom to imagine not only the Milky Way system but also the entire universe as known to us combined in it.

A fireball as big as the Milky Way—perhaps even the whole cosmos! So massive a fireball would fly apart, ejecting matter that we now see as the receding galaxies. That matter would include heavy elements "baked" in the intense heat, elements that eventually condensed into celestial objects such as the Earth. Von Weizsäcker's "fireball" had much in common with Lemaître's "primordial atom." If the big bang could—at least in theory—solve the puzzle of the origin of the elements, then its validity would be greatly strengthened. How could it be tested?

* * *

Anyone who reads Alan Lightman and Roberta Brawer's book *Origins,* a series of interviews with cosmologists, can't help but notice how often the interviewees say that as children they were inspired by the popular writings of George Gamow and Fred Hoyle. As a youngster, I read and learned from the Gamow books, particularly the "Mr. Tompkins" stories. Their protagonist is an imperturbable member of the English bourgeoisie who keeps having bizarre encounters with everyday manifestations of fabulous physics from the macro- and microworlds: for example, bicyclists who peddle so fast that they experience Einsteinian time dilation, and motorcars that, thanks to "quantum tunneling," can drive through walls. Hoyle's books are, of course, legendary, and include *Frontiers of Astronomy* and *The Black Cloud.*

Gamow's genius spurred cosmology toward a test of the validity of the big bang and its role in baking the elements. And it was Hoyle's response to Gamow's proposal that fueled one of the most vociferous and public controversies in cosmology.

Born in Odessa, Russia, in 1904, Gamow became one of the celebrated polymaths of the twentieth century. He happily barreled into—and made substantial contributions to—topics ranging from cosmology and atomic physics to genetics and DNA. After leaving Russia, Gamow worked at universities in Europe, then moved to the United States, becoming a professor at George Washington University. "Ninety percent of Gamow's theories were wrong, and it was easy to recognize that they were wrong," his associate Edward Teller has recalled. "But he didn't mind. He was one of those people who had no particular pride in any of his inventions. He would throw out his latest idea and then treat it as a joke."

Gamow "could ask questions that were ahead of his time," recalls astronomer Vera Rubin, who studied under him and whose husband shared an office at the Johns Hopkins University Applied Physics Laboratory with Gamow's associate, Ralph Alpher. "He had no interest in the details; in many ways he may not have been *competent* to carry out many of the details. . . . He was like a kid." He suffered a tempestuous marriage, drank too much, was an enthusiastic gossip and a legendary practical joker. Amidst all this rambunctiousness, Gamow managed to do some very important science. After World War II, Alpher became acquainted with atomic research at Argonne National Laboratory in Illinois and Brookhaven National Laboratory on Long Island—the place where I did my first particle physics experiments. The atomic data proved to be of great value in his and Gamow's effort to explain the origin of elements formed during the big bang with a theory they developed in collaboration with Robert Herman, also of the Johns Hopkins Applied Physics Laboratory. The research at Argonne and Brookhaven provided two pillars for the theory.

Researchers there fired high-speed neutron beams at metal targets, thereby developing data on the likelihood that a given neutron will pass a nucleus close enough for them to fuse. (This is easier than it is for protons, because neutrons have no electrical charge and, hence, aren't repelled by the protons in the nucleus.) Physicists determined that neutrons in a "free" state (not attached to an atomic nucleus) will spontaneously decay within a few minutes into a proton, an electron, and a neutrino in order to conserve electric charge and other properties.

The stage was set. Gamow proposed that in the beginning, the universe consisted of a primordial substance he called *ylem* (Greek for "primordial matter"). The ylem was an extremely hot (10 billion degrees) gas of neutrons. Because many of the neutrons were "free," they began decaying into protons plus the obligatory electrons and neutrinos. The result was a boiling sea of neutrons and protons. In the intense heat, protons and neutrons began merging into heavier and heavier elements. Some elements with the same numbers of protons had varying numbers of neutrons (these are known as isotopes). In Gamow's view, all the elements in the entire universe formed in this manner during the earliest twenty minutes after the big bang—"in less time than it takes to cook roast duck and potatoes." Because free neutrons were continually turning into protons, after a while the total number of protons was greater than the number of neutrons; hence a lot of protons did not get neutron partners and much of the primordial material remained as protons—the nuclei of hydrogen. Although the hypothesis seemed to account readily for the origin of hydrogen and helium, it worked less well for heavier elements. As the universe expanded and cooled, opportunities and energies required for building heavier nuclei would have diminished. Although he couldn't be certain, Gamow suggested that, nevertheless, the heavier elements might have formed in the aftermath of the big bang.

Much of the detailed calculation of the hypothesis was done by Alpher, forming the basis of his doctoral dissertation, which he submitted in 1948. He and Gamow wrote a paper on the topic, for *Physical Review,* giving Gamow an opportunity to perpetrate one of his jokes. He invited Hans Bethe to be a coauthor of the paper, and thus the "Alpher, Bethe, Gamow" (or alpha, beta, gamma) paper came to be a famed entry in the cosmological literature.

Alpher and Herman wrote a follow-up paper that contained a simple but profound prediction, one that forms the basis of modern cosmology. In order for the protons to fuse into nuclei,

Left to right, Robert Herman, George Gamow, and Ralph Alpher in a composite photograph showing Gamow emerging from a bottle of "ylem," the hypothetical primordial material of the big bang. The montage was Alpher and Herman's joke. RALPH ALPHER AND ROBERT HERMAN

they had to have enough energy to overcome the electrical repulsion of the protons in the target nuclei. This required the early universe to be hot. But if it was too hot, the energetic protons, neutrons, and photons would have blasted apart the nuclei as quickly as they formed. Thus the temperature a few minutes after creation had to lie within a narrow range. Working out the details, they noted that the universe began as an intense fireball of creation, and as it expanded, the radiation—the heat—would not have persisted but become steadily diluted. Beginning with a temperature of many billions of degrees, the universe would gradually have cooled as time passed and space expanded. They calculated that the temperature of today's universe should be about 5 degrees Kelvin (which is the same as 5 degrees Celsius above absolute zero). If the big bang had occurred, as Gamow, Alpher, and Herman predicted, the universe would be bathed in a faint background radiation—an echo of that primordial event—that is the temperature equivalent of 5 degrees Kelvin. Had there been no big bang, no such radiation should be present.

There it was in Alpher and Herman's paper: the prediction of cosmic background radiation, a tangible clue to the big bang. In the 1940's there was no way to detect such a faint afterglow in space, and so Alpher and Herman's prediction remained forgotten for two decades, except by its authors—forgotten like Mendel's historic paper on the genetics of peas.

Gamow was happy to sell his ideas to a mass audience. He told the readers of *Scientific American,*

> To many a reader, the statement that the present chemical constitution of our universe was decided in half an hour five billion years ago will sound nonsensical. But consider a spot of ground on the atomic proving ground in Nevada where an atomic bomb was exploded. . . . Within one microsecond the nuclear reactions generated by the bomb produced a variety of fission products. In 1956, 100 million million microseconds later, the site was still "hot" with the surviving fission products. The ratio of one microsecond to three years is the same as the ratio of half an hour to five billion years! If we can accept a time ratio of this order in the one case, why not in the other?

A persuasive argument, at least to lay readers. He also wrote a book, *The Creation of the Universe,* that sold briskly and included a calculation indicating the cosmic background radiation would be 50 degrees K—not 5 K, the correct figure. As usual, Gamow had made a simple mathematical error.

Across the Atlantic Ocean, in England, Fred Hoyle mocked the Gamow-Alpher-Herman theory—centered on the big bang origin of the universe—and prepared what he hoped would be a devastating response.

Chapter 4

Cosmological Conflict

After dinner on their first evening in the Beach Hotel with the old professor talking about cosmology, and his daughter chatting about art, Mr. Tompkins finally got to his room, collapsed on to the bed, and pulled the blanket over his head. Botticelli and Bondi, Salvador Dalí and Fred Hoyle, Lemaître and La Fontaine got all mixed up in his tired brain, and finally he fell into a deep sleep.

—George Gamow

Mr. Tompkins in Paperback *(1965)*

Without continuous creation, the Universe must evolve toward a dead state in which all the matter is condensed into a vast number of dead stars.

—Fred Hoyle

The Nature of the Universe *(1950)*

Fred Hoyle was born in 1915, the son of a financially insecure textile merchant in the village of Bingley, in northern England. He was three when his mother taught him the multiplication tables. His valley was devoted heavily to the wool industry, so he heard the clatter of machinery from the factories as he walked to school every day. Perhaps he was naturally pugnacious and independent, or perhaps the tense times molded him that way: The First World War was grinding up hundreds of thousands of young men only a few hundred miles away in the trenches of France, and zeppelins were bombing London. Whatever the cause, at an early age he decided he hated school and determined that he would learn about the natural world on his own instead of enduring colorless lectures from drab schoolmasters.

Science was especially exciting at that time because of a new wonder, radio: "This was a great mystery in our village, and there were 20 or 30 people who were wiring up their own little radio receivers." At age ten he began experimenting with chemicals from a home chemistry set. Still, the law required him to attend classes, and there he had some of his first battles with orthodoxy. A teacher struck him when he miscounted the number of petals on a flower, and the young Fred was so enraged that he refused for a while to return to school. He discovered the popular writings of Arthur Eddington, who spurred his interest in cosmology. It is also conceivable that he took note of the aloof, arrogant Eddington's comment that the "notion of a beginning of the present order of Nature is repugnant to me"; perhaps those words unconsciously inspired Hoyle to devote his life to fighting the notion that the cosmos began at a certain point in time, with a big bang. He preferred the view Aristotle held millennia earlier: The universe has always existed, and always will.

A turning point in Hoyle's young life came at age thirteen, when his parents gave him the gift that has changed so many other young lives: a small telescope. They allowed him to stay up all night looking at the stars and planets. As fate would show, Hoyle and Gamow had more in common than the fact that each had received a telescope in his thirteenth year. Each was a fighter, intellectually speaking, and each exploded with far more ideas than could ever be true. After working on radar in England during the Second World War, Hoyle became an astronomy professor at Cambridge University. He also began delivering talks about astronomy on BBC radio and writing popular articles and books. Like Gamow, Hoyle was becoming a highly visible interpreter of science to laypeople.

During one of his popular radio broadcasts in 1950 Hoyle coined the phrase *big bang* as a description of Gamow's repugnant (to Hoyle) theory. Hoyle had meant the term to be derogatory, but it was so compelling, so stirring of the imagination, that it stuck, but without the negative overtones. Hoyle became the most visible proponent of an alternative theory to the big bang, known as the steady state theory. The struggle for intellectual supremacy between these two theories dominated cosmology for almost two decades.

One night in 1946, Hoyle went to see a movie called *The Dead of Night*, in the company of two Cambridge colleagues, Thomas Gold and Hermann Bondi. The plot was a series of ghost stories with an unusual ending: The last scene of the film is the same as the first. It was a cyclical story, continuous, without end. According to legend, this cinematic twist inspired Gold to propose that the cosmos was cyclical, too. Big bang theory implied that as space expanded and the galaxies spread outward, the cosmos would gradually dissipate like a cloud. Gold rejected that notion and argued that the universe would never dissipate because it was not only constantly generating new space for the expansion, it was also generating new *matter* to replace the old galaxies! Where did the new matter come from? From nothing! Literally, the universe was constantly spawning new atoms from the vacuum. These new atoms gathered into galaxies, which continued expanding outward.

The universe was, in short, in a steady state—always remaining essentially the same as it is today. It was homogeneous not only in space (as had been generally held for years) but also in time. It would look the same 10 billion years ago, or in the future, as it does today. The three theorists liked the idea, not least because it avoided one of the main headaches of big bang theory: its positing of a beginning of space and time, a beginning beyond which the human mind could never fathom.

They published separate papers on the subject in 1948—Bondi and Gold together, Hoyle alone. Hoyle's approach was more mathematical, Bondi and Gold's more philosophical. But Hoyle took unmistakable philosophical pleasure in the theory's reassuring implications: The universe would never die. Until then, its doom had appeared inevitable. According to nineteenth-century thermodynamics, closed systems gradually decline into disorder (their entropy increases), and such a fate seemed to await the universe. Physicists began to think of the cosmos as a great mechanism that, like any machine, must eventually run down. The shining stars would eventually burn out; the whirling planets would eventually crash into their parent suns. This melancholy view found literary expression in the closing pages of H. G. Wells's 1895 science-fiction tale *The Time Machine,* about a man who travels millions of years into the future. Standing by an ocean on the near-lifeless future Earth, he observes that the world is slowing and growing darker. The planets have stopped rotating and are revolving closer and closer to the dying Sun:

> . . . a steady twilight brooded over the Earth. . . . All trace of the moon had vanished. The circling of the stars, growing slower and slower, had given place to creeping points of light . . . the sun, red and very large, [had] halted motionless upon the horizon, a vast dome glowing with a dull heat. . . . The rocks about me were of a harsh reddish colour, and all the trace of life that I could see at first was the intensely green vegetation . . . the same rich green that one sees on forest moss or on the lichen in caves: plants which like these grow in a perpetual twilight. . . . I cannot convey the sense of abominable desolation that hung over the world.

A similar gloom was expressed by the philosopher Bertrand Russell, in 1923:

> All the labors of the ages, all the devotion, all the inspiration, all the noonday brightness of human genius, are destined to extinction in the vast death of the solar system; and the whole temple of Man's achievement must inevitably be buried beneath the debris of a universe in ruins. . . .

If steady state theory were true, Hoyle said, then the bleak visions of Russell and the nineteenth-century "heat death" visionaries need never come true. "Without continuous creation," he wrote in 1950, "the Universe must evolve toward a dead state in which all the matter is condensed into a vast number of dead stars. . . . With continuous creation, on the other hand, the Uni-

Fred Hoyle, champion of the steady state theory of the universe.

verse has an infinite future in which all its present very large-scale features will be preserved." To the "steady statesmen," the continuous creation of matter guaranteed the universe was homogeneous not only in space—as maintained by the so-called cosmological principle—but also in time. Hence, Hoyle and his colleagues called their idea the *perfect* cosmological principle. An astronomer would see essentially the same type of universe at any time, either billions of years in the past or billions of years in the future.

The appeal of the steady state theory was its intellectual power: It was much more tractable to scientific analysis. The big bang model posits a beginning of the universe, when conditions were far different from those known today, and hence beyond what physicists could explain (at least, with knowledge as it was in the 1940's and 1950's). In the steady state model, there was no such black hole of ignorance. In their 1948 paper, Bondi and Gold argued that the steady state universe was "compelling, for it is only in such a universe that the laws of physics are constant." Physicists felt more at ease with a theory whose proposals they could calculate. For these reasons the steady state theory became popular in the 1950's and early 1960's and was considered just as plausible as the big bang.

True, steady state theory was based on a seemingly indigestible idea: the routine creation of matter from nothing.* This "may seem a very strange idea and I agree that it is," Hoyle said, "but in science it does not matter how strange an idea may seem so long as it works—that is to say, so long as the idea can be expressed in a precise form and so long as its consequences are found to be in agreement with observation." He suggested matter might be continuously created by a hypothetical "creation field," akin to a gravitational or electromagnetic field. The rate of matter creation was set by the rate at which galaxies receded; the faster they receded, the greater the rate of matter creation. In fact, the creation of new matter caused the cosmos to expand. The "inexorable introduction of new units of creation [forced] the others apart, much as the introduction of new guests into a cocktail party forces earlier guests to move outwards from the initial gathering point."

How fast was new matter created? As fast as necessary to replace matter lost as galaxies receded into deep space. This amounts to one hydrogen atom every second in a cube with a 160-kilometer (100-mile) side; or stated somewhat differently: that about a quarter of a million hydrogen atoms originate every second in a volume of space equal to the volume of the Earth; or about one atom every century in a volume equal to the Empire State Building. For the cosmos as a whole, that was a magnificent

*The idea of generating something from nothing wasn't new: Robert Millikan had claimed cosmic rays were photons generated by the continuous creation of matter in space—creation that, he argued, showed "God is still on the job." And in his book, *Astronomy and Cosmogony,* astrophysicist James Jeans had suggested the "centres of the nebulae are of the nature of 'singular points' at which matter is poured into our universe from some other, and entirely extraneous, spatial dimension, so that, to a denizen of our universe, they appear as points at which matter is being continuously created." Ironically, in response to the common question "What came *before* the big bang?" Edward Tryon and other theorists have recently used quantum theory to argue that our cosmos literally formed from nothing.

rate of production: In tons, the total amount generated per second was at least one followed by thirty-two zeros. Such a rate threatened to violate the principle of the conservation of mass and energy. The steady state theorists admitted that it posed a theoretical conundrum, but insisted that it was easier to envision the slow, steady creation of matter from nothing than the instant creation of all matter portrayed by the big bang theorists.

Steady state theory's one clear advantage over the big bang—that it did not have to explain what happened before creation of the universe—had broad emotional appeal. Eddington's early request that any cosmic beginning not be "unaesthetically abrupt" reflected his deep conviction that scientific explanations shouldn't have a "presto" quality, like a magician pulling a rabbit from a hat. To early critics, the big bang theory suffered from exactly that problem; it implied that all the features of our cosmos—its types of forces, physical constants, etc.—were arbitrarily set from the beginning; or, in other words, were special "initial conditions." If this was true, then (critics charged) their origin was forever beyond scientific understanding.

Hoyle characterized this approach as like "primitive peoples . . . who in attempting to explain the local behavior of the physical world are obliged in their ignorance of the laws of physics to have recourse to arbitrary starting conditions . . . by postulating the existence of gods, gods of the sea who determine the arbitrary starting conditions that control the motion of the sea, gods of the mountains, gods of the forests. . . . The aim in science is not to build a theory that is so hedged in with protective conditions that nobody can get at it." Theorists never like to be told that something is beyond their potential understanding. Likewise, to Gold, big bang theory was circular in its reasoning because, he said, it held that "the universe is what it is because it was what it was." Or as one steady statesman joked, the big bang was as awkward as a girl jumping out of a cake.

At times the debate between the big bang and steady state theorists resembled more a confrontation between different schools of philosophy than different schools of cosmology. At times it became quite emotional with the big bang advocates responding in kind to the taunts of Hoyle and others. In his book *Mr. Tompkins in Paperback* (1965), Gamow ridiculed the steady statesmen—whose intellectual home was Cambridge, England—by noting that their ideal of an unchanging cosmos "is in accordance with the good old principle of the British Empire to preserve the status quo in the world."

Ultimately, neither jokes nor philosophy would decide the war between the big bang and steady state theories. Instead, these theories, like all theories in science, would be validated by how their predictions matched observations. The steady state and big bang theories made several distinct predictions about the age of the cosmos, the abundances of various elements, the distribution of matter across space and time, and the relic radiation from the primeval fireball. Let's look at each prediction.

1. The Age of the Cosmos

Based on the rate of galactic recession, Gamow and his colleagues calculated the age of the universe to be 1.8 billion years. The steady statesmen thought the cosmos was infinitely old.

At first, the big bang theory seemed to founder on this crucial point. Radioisotope dating of rocks indicated Earth was 4 or 5 billion years old. If Gamow's calculations were right, that meant the planet was far older than the universe—an obvious absurdity. For a time, this age discrepancy was a major source of encouragement to the steady state theorists. The discrepancy vanished in the 1950's, however, when the German astronomer Walter Baade (1893–1960) showed that Hubble had underestimated the distance to the galaxies, thus underestimating the age of the universe. The cosmos was in fact far older—probably between 10 and 20 billion years, which left plenty of time for Earth to form.

2. The Abundances of Elements

Based on their theory for the origin of elements, Gamow, Alpher, and Herman predicted that about three fourths of cosmic matter is hydrogen and about one fourth is helium. All heavier elements constitute less than 1 percent. Also, they said, the elements should be distributed fairly evenly across the universe because the big bang occurred everywhere. In contrast, Hoyle and his supporters predicted that elemental abundances would vary across space if elements were formed in local celestial furnaces—stars—as proposed by the steady state theory.

As is so often the case in debates, both sides were partly right. Gamow and his colleagues had been confident about the big bang's ability to produce the observed abundance of hydrogen and helium, but were less so about the same processes generating the heavier elements. In 1949–50 Enrico Fermi and Anthony Turkevich at the University of Chicago showed that Gamow's uncertainty was justified: Elements heavier than hydrogen and helium could not have been manufactured by the big bang. Their experiments showed that isotopes with mass numbers of 5 and 8 were unstable and wouldn't have lasted long enough to absorb more protons and neutrons and, thereby, to aggregate into even heavier elements. The big bang advocates were therefore able to explain the origin of light elements (hydrogen and helium), but not heavy ones (everything heavier than helium).

The steady state theorists faced the opposite bind: Nuclear processes within stars could produce the heavier elements, but not hydrogen and helium. By the mid-1960's Hoyle admitted he was stumped when it came to providing a simple steady state explanation for the abundance of helium. In contrast, the high heat and density of the early big bang universe clearly could account for the abundance of helium.

Hoyle made a major contribution to understanding the origin of heavy elements, in calculations he published in the 1940's and 1950's. He, William A. Fowler, and Geoffrey and Margaret Burbidge showed how heavy elements could have been "cooked" inside stars, with hydrogen and helium as raw material. Eventually these stars exploded and ejected the debris into the galaxy. After millions of years, the debris condensed into new stars and planets. Their work revolutionized our understanding of elemental abundances and nowadays is the generally accepted explanation for the origin of heavy elements. Most of your body consists of atoms brewed inside a star that later detonated. We are all literally made of stardust.

The origin of the elements is therefore a two-stage process. Hydrogen and helium are produced in the early stages of a primordial creation event (the big bang); heavy elements are subsequently manufactured by nuclear processes within stars. The big bang theory, revised by the work of Hoyle et al., is consistent with this picture; the steady state theory is not.

3. The Distribution of Matter Across Space and Time

Steady statesmen predicted that astronomers should see (a) galaxies of widely varying ages, from extremely old to extremely young, and (b) no significant change in the density or type of galaxies across space or time. The reason, they said, is that the cosmos is eternal and that matter is always being created; hence new galaxies are always forming to replace the old ones that retreat into space.

In contrast, big bang theorists predicted that astronomers should see ever larger numbers of galaxies per unit volume as they looked deeper and deeper into space and, thereby, into time (that is, closer to the moment of the big bang). That's because—they held—all matter was created in the big bang; therefore, the density of the cosmos should decline as it expands. Also, all galaxies formed soon after the big bang; hence there should be no young galaxies. And by looking farther out and farther back in time, big bang theorists predicted, astronomers should see the evolution of galaxies.

The next prediction is the most famous of all, and it ultimately settled the cosmological conflict.

4. The Temperature of Space

In their 1948 paper, Ralph Alpher and Robert Herman made their now-famous—but long-forgotten—prediction of cosmic background radiation. A big bang universe would have begun with intense heat—radiation—and as the universe expanded the radiation per unit volume would have been steadily diluted. This ubiquitous soup of radiation would therefore have cooled as time passed. By now, they calculated, this radiation afterglow would have cooled to a temperature of about 5 degrees Kelvin.*

*Five degrees Kelvin, or 5 K, is temperature in degrees centigrade above absolute zero. The units are named after Lord Kelvin, a pioneer in heat research. Absolute zero is the temperature of an object when all its available heat is removed. It is about −273 degrees centigrade or 490 degrees Fahrenheit (−460°F) below the freezing point of water.

They suggested that someone should try to detect it. Sixteen years passed after Alpher and Herman's prediction before anyone tried to detect the residual radiation. In contrast, the steady state model made no prediction about the temperature of the universe.

> *It is a suspicious feature of the explosion [big bang] theory that no obvious relics of a superdense state of the Universe can be found.*
>
> —*Fred Hoyle*
> Frontiers of Astronomy *(1955)*

The first serious threat to the steady state theory involved our third prediction: the distribution of matter across space and time. Ironically, the threat came not from Gamow and his colleagues across the Atlantic Ocean but from one of Hoyle's colleagues at Cambridge—radio astronomer Martin Ryle.

Radio astronomy transformed modern astronomy as Galileo's telescope had transformed postmedieval astronomy. Whereas traditional telescopes studied a narrow slice of the electromagnetic spectrum—the optical realm, which we perceive with our eyes—radio astronomy began the opening of the rest of the spectrum, such as the radio, microwave, infrared, and (in time) gamma ray frequencies. As a result astronomers have discovered weird, incredibly energetic phenomena including quasars, neutron stars, and (perhaps) black holes. Radio telescopes also enabled them to look much deeper into space and, hence, much farther back in time—and, therefore, provided the ability to test cosmological theories, especially predictions 3 and 4.

Radio astronomy emerged by accident. In the 1930's, Karl Jansky of the Bell Telephone Laboratories noticed curious radio emissions from the center of our galaxy, from a region in the direction of the constellation Sagittarius. He built an array of antennas to create a crude map of the radio emissions. Soon afterward, the American amateur astronomer Grote Reber built his own device to listen to celestial radio signals. What really made radio astronomy a going concern was World War II: The development of radar made astronomers aware they could use nonoptical parts of the electromagnetic spectrum to probe space. For example, in 1946 a radar signal was bounced off the Moon. In following years astronomers at the Jodrell Bank radio telescope in England and others elsewhere mapped radio sources in the sky, some of which were surprisingly intense. They were dubbed "radio stars" and "radio galaxies."

Back then, most radio telescopes looked less like the yawning dishes of today and more like fields covered with wire-mesh grids and poles. These static structures were too big to point at this or that celestial object. Instead, they passively "mapped" the radio sky as Earth rotated. It was a tedious process. Would modern astronomy have emerged had Galileo been forced to sit all evening, looking in one direction through his optical telescope, while waiting for the Moon to pass in front of it?

A physics graduate of Oxford University, Ryle had worked on radar during the war and entered radio astronomy afterward. In 1950, he and his colleagues found evidence of radio emissions from four nearby galaxies, including Andromeda. A few years later he mapped the distribution of radio sources across the northern sky using a 3.7-meter-wavelength array (resembling futuristic power lines) covering an acre just outside Cambridge.

His results stunned the steady statesmen: Radio galaxies were not distributed evenly across space. The more distant they were, the more abundant they were. That meant the density of radio galaxies was greater in the past (because the farther one looks into space, the deeper one sees into time). This contradicted the steady state prediction that matter is evenly distributed across space and time. Controversy followed. An Australian radio observatory reported radio-source measurements less clear-cut than Ryle's. But in the main, Ryle's original view has been upheld: The primeval universe was denser than the present universe. Hence the cosmos isn't homogeneous in space and time as purported by the perfect cosmological principle of Hoyle, Bondi, and Gold.

An even more celebrated example of the nonuniformity of matter was discovered in the early 1960's: quasars. These "quasi-stellar objects" had extremely high redshifts (some have wavelength changes of a factor of three or four). Hence, according to the Hubble law, the quasars formed an extremely long time ago, almost as long ago as the big bang. Most important, they were completely different from anything seen in the modern universe; they emitted astonishing amounts of radiation from relatively tiny areas. Quasars are now thought to be very active nuclei/centers of young galaxies. In short, quasars portrayed a primeval universe far different from the current one—and, therefore, far different from the never-changing universe of the steady state theory.

Ryle's findings, with their threat to Hoyle's steady state theory, delighted Gamow. His wife, Barbara Gamow, wrote a poem that he published in one of his books:

"Your years of toil,"
Said Ryle to Hoyle,
"Are wasted years, believe me.
The steady state
Is out of date.
Unless my eyes deceive me,
My telescope
Has dashed your hope;
Your tenets are refuted.
Let me be terse:
Our universe
Grows daily more diluted!"
Said Hoyle, "You quote
Lemaître, I note,
And Gamow. Well, forget them!
That errant gang
And their Big Bang—
Why aid them, and abet them?
You see, my friend,
It has no end
And there was no beginning,
As Bondi, Gold,
And I will hold
Until our hair is thinning!"

As the case against steady state theory mounted, Hoyle vacillated. He conceded in *Nature* in 1965 that "the indication of the radio counts is that the universe was more dense in the past than it is today" and that quasars indicated "the universe has expanded from a state of higher density. . . . [It] seems likely that the [steady state] idea will now have to be discarded, at any rate in the form it has become widely known." In later years, however, he drifted back to a modified version of steady state. To this day, he insists that some form of continuous creation theory is preferable to the big bang. In that sense, Mrs. Gamow was prophetic: I can tell you Hoyle's hair is silver-white, although it does not seem to have thinned noticeably.

I met Hoyle on a number of occasions, for example, in December 1991 at a conference on "Observational Cosmology" in Durham, England. He had the room next to mine during the conference; we often talked and ate together. I found him sharp and pugnacious. When one of the participants tried to give him credit for inventing a precursor of inflation theory—namely, his famous "C-field"—Hoyle wanted nothing to do with it. To him, a universe with a beginning (which inflation theory endorsed) was more of an anathema than not getting scientific credit.

Nowadays Hoyle has gained perspective and talks more gently about his past battles. But occasionally, bitterness seeps through, as indicated by remarks he made at a conference in Italy in 1988:

> From about 1955 onwards Ryle had the idea that by counting radio sources as a function of their fluxes he could disprove the steady-state theory. His programme, which he pursued relentlessly over the years, does not seem to have been directed towards any other end. There was no question of establishing the correct cosmology, but only of disproving the views of a colleague in the same university, a situation which I have never felt to have deserved the plaudits which the scientific world showered upon Ryle. . . . I am left wondering why anybody should have believed Ryle. . . . The only reasonable explanation I can offer is that Ryle's claims were widely considered to be culturally desirable.

Hoyle didn't clarify what he meant by "culturally desirable." But he wasn't the first cosmologist in history who accused his critics of being swayed by nonscientific motives. Likewise, other critics of the big bang theory claim it's popular because (in the view of some Christian apologists) it resembles the creation as depicted on the first page of Genesis. They also enjoy pointing out that the first big bang theorist, Lemaître, was a priest. "[O]ur contemporary myths like to garb themselves in scientific dress in pretense of great respectability," charges a distinguished big bang foe, Hannes Alfvén, a Swedish Nobel laureate from the University of California at San Diego.* At the same time, big bang theorists haven't been entirely pure when it comes to imputing the motives of their foes. A prominent big bang cosmologist, Edward R. Harrison, has even suggested that Eddington opposed theories of cosmic beginnings and ends because he felt uncomfortable with issues of birth and death, being "a bachelor who lived with his sister."

All in all, steady state theory was on the brink of a cliff by the mid-1960's. Only a nudge was needed to push it over.

In August 1959, when I was fourteen years old, the United States launched a satellite called *Echo 1.* It was a giant balloon as tall as an office building, and so highly reflective that you could easily see it from Earth. The night after it entered orbit, my parents and I went out into our yard and saw it passing overhead. Off that bright speck of light, radio engineers bounced a spoken message from President Eisenhower; their goal was to determine whether communications satellites, first proposed in 1945 by Arthur C. Clarke, were feasible.

A peculiar kind of radio antenna had been built to do experiments with *Echo 1.* Located in Holmdel, New Jersey, the antenna resembled a large horn—large so it could detect the faint signals—and was operated by AT&T's famous Bell Laboratories, the birthplace of the transistor and other inventions. Its developers didn't realize their horn would also be able to detect a different form of radiation, a form that had been predicted almost two

*Lemaître insisted his theory had nothing to do with religion. In fact, he was upset when, in 1951, Pope Pius XII cited the big bang as evidence of divine creation.

decades earlier, then forgotten: the cosmic background radiation.

As so often happens with major scientific advances, several individuals had come close to, but for various reasons never crossed, the threshold of discovery. As far back as the late 1940's, scientists had recorded unusual background radiation from the sky, but they were unable to determine its source. They generally concluded it was either stray noise or an instrumental error; no one seems to have suspected it might come from the cosmos. In 1946 Robert Dicke invented the differential microwave radiometer, a sensitive instrument for detecting patterns of cosmic radiation. He used it to scan the sky, and concluded that whatever background radiation existed, it was cooler than 20 degrees Kelvin. At that time he was simply looking for deep-sky radiation of any kind, not specifically for cosmological radiation such as the relic glow of the big bang.

In the spring of 1964, Arno Penzias and Robert Wilson, researchers at Bell Labs, were using the Holmdel horn to measure noise levels that might contaminate communications with *Echo 1*. Unexpectedly, they picked up a persistent microwave radiation (at a wavelength of 7.35 centimeters), which had a temperature equivalent of about 3.5 degrees Kelvin, and were unable to eliminate it from their system. The signal was coming from all directions of the sky. At one point they speculated it might be caused by what they primly called a "white dielectric material"—pigeon droppings. Or perhaps the signal was caused by heat from the pigeons' bodies (the antenna was that sensitive!). They cleaned out the droppings, but the signal persisted. A colleague, radio astronomer Bernard Burke, of the Massachusetts Institute of Technology, suggested the signal might be a cosmological phenomenon and advised them to contact the astrophysicists Robert Dicke and James Peebles, who were just down the road at Princeton.

About that time, Peebles had been going around giving talks about how light elements (such as hydrogen and helium) might have been manufactured in the early universe. He was unaware that in the late 1940's, Gamow, Alpher, and Herman had proposed that these elements (plus the heavy ones) were formed by the big bang, and that the result would be a cosmic bath of primordial radiation now cooled to a residual temperature of

Robert Wilson (*left*) and Arno Penzias; behind them is the Holmdel, N.J., antenna with which they unexpectedly discovered cosmic background radiation. AT&T ARCHIVES

about 5 degrees K. Dicke and Peebles independently calculated that if the big bang theory was correct, then the primordial radiation should exist and should be detectable. Initially, they calculated the background radiation should be about 10 degrees K. They planned an experiment to detect it.

One day Dicke held a lunch meeting in his office with colleagues working on the cosmic background radiation experiment—Peebles, David Wilkinson, and Peter Roll. The phone rang and Dicke answered. Soon the others heard him using words such as "background radiation" and "3 K"; their ears perked up. Eventually Dicke said good-bye, put down the phone, turned to his associates, and said: "We've been scooped." The call had come from Penzias and Wilson. They had described the odd radiation and had no idea what it was; did Dicke? He did.

As a result, the two teams agreed to publish two papers jointly in the same issue of *Astrophysical Journal Letters*. The Penzias-Wilson paper was entitled "A Measurement of Excess Antenna Temperature at 4080 Megacycles per Second"; it was modestly worded and described their observations, nothing more. But it included this innocent-looking line: "A possible explanation for the observed excess noise temperature is the one given by Dicke, Peebles, Roll and Wilkinson (1965) in a companion letter in this issue."

That "possible explanation" was the bombshell: the proposal that the Holmdel antenna had detected leftover radiation from the creation of the universe—the big bang. The Dicke et al. paper described how the big bang could have generated primordial radiation that, by now, would have cooled to a particular temperature. The result was a front-page story in the May 21, 1965, *New York Times*, announcing that compelling evidence for the big bang had finally been found. Wilson indicated he hadn't appreciated the importance of their discovery until he saw the news on the front of the *Times*.

In technical jargon, Penzias and Wilson's discovery sounds dull: They had detected isotropic "blackbody" radiation with a temperature of about 3 degrees K at a microwave wavelength of about 7.35 centimeters—that is, the distance between each wave crest was 7.35 centimeters. Also, the waves were arriving at the rate of 4.08 billion per second (that is, a frequency of 4080 megahertz). (The speed of light is very fast, 2,977,000,000 centimeters per second.) But the real smoking guns were the isotropic and blackbody nature of the radiation. *Isotropic* means that the radiation at the wavelength was equally intense all over the visible sky. That's exactly what one would expect if relic radiation came from the big bang: It occurred everywhere simultaneously, hence its afterglow should be uniform across the heavens.

Blackbody radiation arises whenever particles collide with each other very rapidly in thermal equilibrium. Given the intensity of such collisions in the posited big bang, a huge flux of such radiation should have been produced early on, and should still be present throughout the universe. Blackbody radiation is readily identified by its "Planckian" spectrum.*

Most people are familiar with the way a glass prism (or even a raindrop) can fragment visible light into the colors of the spectrum. Each color represents a different wavelength or frequency of light, and the spectrum we see has a characteristic distribution of intensity of radiation at each frequency (color). Likewise, radio astronomers can analyze a celestial signal by dividing it into different wavelengths, or frequencies. The "spectral curve" shows how the radiation intensity varies with frequency. Blackbody radiation has a very characteristic spectrum, which resembles a slightly lopsided hump on a camel. (See figure on page 86.)

According to big bang theory, the universe was in thermal

*So named for Max Planck, who in 1900 pioneered quantum theory, which explains the intensity versus frequency curve (spectrum) for blackbody radiation. Quantum theory was one of the greatest revolutions in the history of science, and stemmed directly from blackbody radiation research. Before developing it, Planck and other physicists had been troubled by a physical puzzle known as the "ultraviolet catastrophe," which was, in a sense, a down-to-Earth version of Olbers's paradox. Nineteenth-century physicists had reasoned that since a blackbody perfectly absorbs and emits all frequencies, most of its radiation should be at high frequencies, i.e., the ultraviolet range and higher; that's because the number of states available at high frequencies is much greater than the number at low frequencies. So why don't blackbodies emit much more high-frequency radiation than they do? After wrestling with the question for years, Planck bit the bullet and decided only a radical answer would suffice—an answer that, indeed, defied "common sense." In a speech and paper in 1900, he proposed that energy isn't emitted in a continuous, infinitely divisible fashion but, rather, in packets called quanta. The release of higher-frequency light quanta can only occur by correspondingly high-energy inputs of energy. Hence a blackbody tends to emit a far higher amount of low-frequency and thus low-energy quanta than high-frequency and thus high-energy quanta. This idea was as revolutionary as Einstein's theories of relativity because it showed that matter and energy are, ultimately, discrete rather than continuous.

equilibrium during its early stages. A searing light pervaded all locations and traveled in all directions, with the characteristics of a blackbody at very high temperature. Early on, the temperature was in the trillions of degrees. As space expanded, the wavelengths of light stretched out too. You could compare this to vertical lines marked on a rubber band. As you stretch the band, the lines spread apart. Likewise, the expansion of space stretches the wavelengths shifting the "trademark" blackbody spectrum to that of a lower temperature. Blue light shifted to the cooler red region and so on. Hence the universe cooled.

For about three hundred thousand years the residual blackbody radiation was energetic enough to prevent electrons and protons from binding together: Each time an electron went into orbit around a proton, it was knocked away by an energetic photon (quantum of light) from the hot radiation. The universe was effectively an opaque soup of particles and photons, inextricably associated in high-energy interaction. But as time passed, the universe cooled further. At three hundred thousand years things were so cool that photons lacked the energy to dislodge electrons from protons. This had two results: Protons and electrons began to associate and remain stable, in the form of hydrogen nuclei; and the photons were now free to flux where they might. With this rapid decoupling of radiation and matter the universe became transparent, and radiation streamed in all directions—to course through time as the cosmic background radiation we experience still, a perpetual reminder of our fiery birth.

People sometimes ask why we can still see the cosmic background radiation. Why didn't its light just rush past us and disappear into space, like a flash of light from a distant explosion? This reflects a common and understandable confusion about the true nature of the big bang. The classical big bang didn't happen at a specific place within an infinite void; rather, it happened everywhere because it was everything. There was "nothing"—not even empty space—outside of it. Hence the radiation is everywhere and goes in all directions, and will continue to do so as long as the universe exists, as it travels unhindered (save for extremely rare interactions of some of the light with gas, dust, and galactic matter). Every year we see cosmic background light that started toward us at the same time as the previous year's but that had one light-year farther to travel. Because of the extra distance the radiation had to travel, it is slightly more redshifted—about one part on 15 billion (the universe is roughly 15 billion years old). That change is much too small to observe.

Penzias and Wilson's discovery of the cosmic background radiation was a fatal blow to the steady state theory. In 1967, after valiant resistance, steady state theorist Dennis Sciama threw up his arms and admitted the big bang theorists had won. "For me," he lamented, "the loss of the steady-state theory has been a cause of great sadness. The steady-state theory has a sweep and beauty that for some unaccountable reason the architect of the universe appears to have overlooked. The universe is in fact a botched job, but I suppose we shall have to make the best of it."

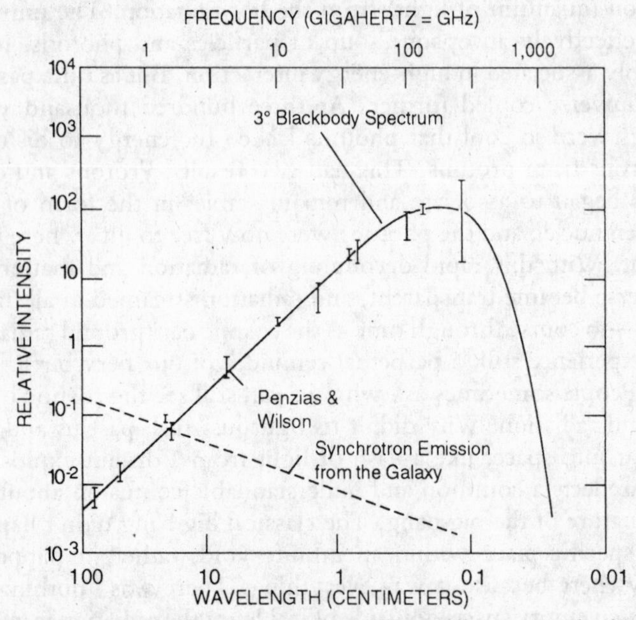

Spectrum for a blackbody of temperature of about 3 degrees K. The discovery observation labeled "Penzias & Wilson" is shown as a bar but represents the probable intensity given the experimental errors. The other bars show other observations up until the mid-1970's. The dashed line shows the intensity interfacing radiation from our own galaxy. The frequency is labeled in gigahertz, a unit representing one billion cycles per second, or one thousand megacycles per second.

Chapter 5

In Search of Antiworlds

Perils of Modern Living

Well up beyond the tropostrata
There is a region stark and stellar
Where, on a streak of anti-matter,
*Lived Dr. Edward Anti-Teller.**

Remote from Fusion's origin,
He lived unguessed and unawares
With all his anti-kith and kin
And kept macassars† *on his chairs.*

One morning, idling by the sea,
He spied a tin of monstrous girth
That bore three letters: A.E.C.‡
Out stepped a visitor from Earth.

Then, shouting gladly o'er the sands,
Met two who in their alien ways
Were like as lentils. Their right hands
Clasped, and the rest was gamma rays.
—*H.P.F.*
The New Yorker *(November 10, 1956)*

*Physicist Edward Teller is best known as the father of the American hydrogen bomb.

†Antimacassars are covers to protect the backs or arms of furniture. Anti-antimatter is matter.

‡In 1956 the Atomic Energy Commission (A.E.C.) had responsibility for the development of atomic energy. Its responsibilities have since been passed to the Nuclear Regulatory Commission and the Department of Energy.

The task: to lift two tons of delicate and expensive equipment high into the stratosphere, and then return it safely to Earth, so that tens of thousands of scientific measurements can be analyzed in the comfort of the laboratory.

The means: giant helium-filled balloons, unruly leviathans that shake and crack loudly in the gentlest of breezes and seem constantly to threaten destruction to equipment and experimenters alike, and yet are as fragile as the thin film of plastic from which they are constructed.

I had known that scientific ballooning would be an adventure, which was one reason why I had become involved. It turned out to be much more technically difficult—and much more romantic—than I ever imagined. It also, in the end, broke my heart—or at least dampened my enthusiasm for this brand of scientific adventure.

As with a space launch, preparing for a balloon flight demands that everything be right—everything checked and double-checked. The balloon must be properly stowed, to avoid disastrous entanglement on liftoff. Every piece of equipment must be working perfectly—which in our case included maintaining equipment at low temperature and high vacuum—for there is little or no way to fix things once aloft. After the mission is finished, the gondola is safely parachuted to Earth—in theory, anyway. In reality, the release mechanism may not work. If not, then hundreds of thousands of dollars of scientific instrumentation ends up as so much scrap. And the weather has to be more than just fair; it has to be perfect, otherwise the gondola may disappear into oblivion as winds carry the balloon out of accessible range, and sometimes into the deep ocean.

Something in my nature responds to these kinds of challenges, and the joy of my first launches—beginning in 1971—will always be with me. One of my responsibilities was to charge up a superconducting magnet, which was a key component of the equipment we were launching. The procedure required the magnet to be cooled with liquid helium, and there was an ever-present danger of explosion if the cooling and charging slipped out of balance. I took some pride in having the required steady nerves and skill to do this—despite some nasty setbacks—while others preferred not to be close.* As team field leader I insisted on waiting until the weather was ideal, with only the slightest whisper of wind to endanger the launch or the subsequent landing. For this obsession the local crew christened us the California chickens—someone even fashioned a papier-mâché psychedelic chicken and taped it to the gondola so it could go along for the ride.

The slow, majestic rise of the balloon, once launched, produces a surge of triumph—and relief—at a phalanx of physically and mentally demanding tasks successfully accomplished. For a while you watch as the balloon becomes an ever-diminishing part of the (usually) late-afternoon sky, the low sun illuminating the reflective fabric so that it looks like a bright planet. Soon you are left with nothing but the sounds of the night, your hopes for the mission, and many fears at the awful prospect of something going wrong.

The lure of cosmology had brought me to launching balloons. Beginning in 1966, I became a particle physicist doing my graduate work at the Massachusetts Institute of Technology, where I had the great fortune to have Dave Frisch, Steven Weinberg, Victor (Vicki) Weisskopf, and many others as teachers and examples. The science satisfied my need to be searching for something fundamental in nature—particle physics and cosmology are about as fundamental as you can get, in their scrutiny of the nature of the universe and its origin. A shift was going on in particle physics as I was passing through graduate school, a consequence of the ever-increasing size of the basic equipment: the particle accelerators. For the final experiment I did as a particle physicist, I was a member of a team of five. I knew that the next team would be twenty-five, and that the number would only go on increasing. I was twenty-five years old at the time and would be a junior member of the effort. I realized that in such an environment I would have less opportunity to have an individual impact.

In cosmology, meanwhile, teams were small, and the endeavor itself was on the brink of becoming a modern science.

*Automation has since arrived. Now computer-controlled devices do it.

Traditionally, cosmology has been a mélange of other disciplines, mainly astronomy and metaphysics. I realized that if it imported the techniques of the particle physics approach—organized teams, high-technology data collection, and computer-based analysis—cosmology could become powerful in its own right, and I wanted to be a part of that. I left MIT in 1970 and went to the University of California at Berkeley, to work with Luis Alvarez of the Lawrence Berkeley Laboratory. The lab, a distinguished federally funded facility that has employed a long list of Nobel laureates, is nestled in the woods of the Berkeley Hills, where deer sometimes startle motorists and the lights of San Francisco shimmer across the bay like a nearby galaxy. The lab's three thousand employees are dedicated to a wide range of ground-breaking disciplines from astrophysics to gene mapping to energy conservation, all of it nonmilitary.*

My scientific colleagues there included Andy Buffington, Larry H. Smith, and Mike Wahlig; we were soon joined by Charles Orth. Alvarez, or Luie as he was known, was the leader of the project. To all of us, Luie was both an inspiration and a source of some terror. To the public he is best known for helping develop the theory that the dinosaurs were blasted into extinction when a giant asteroid or comet collided with Earth 65 million years ago. But Luie was also a legendary figure in physics—and the winner of the 1968 Nobel Prize in physics. He was one of the most diverse, ornery, and brilliant humans one could meet, and inexorably had a major influence on our careers and our ideas of how to do scientific research.

Born in 1912, the son of famed newspaper medical columnist Dr. Walter Alvarez of the Mayo Clinic, Luie worked on the Manhattan Project and made fundamental discoveries in particle physics. In addition to shaking up the field of paleontology with his wildly unorthodox (and now widely accepted) idea about dinosaur extinction, Luie investigated a variety of other exotic topics, ranging from the Kennedy assassination (concluding the

*That surprises some visitors, who mistake LBL for Lawrence Livermore National Laboratory, a nuclear weapons and research facility east of San Francisco. The main thing the two labs have in common is "Lawrence" in their names, after LBL founder Ernest O. Lawrence, a pioneer of the particle accelerator and nuclear physics.

"one bullet, one gunman" theory was sound) to the secrets of an Egyptian pyramid (he "X-rayed" the pyramid with cosmic rays). Luie was a demanding mentor who didn't suffer fools gladly. Every Monday night he invited the group to his home, where someone had to speak about a particular scientific topic. One former colleague recalls how, during his first talk there, he showed his first viewgraph and Luie roared, "Terrible, terrible, terrible!" The viewgraph had five mistakes, Luie charged, and he proceeded to point them out. Such onslaughts were great training for a scientific career. "Now I work for a corporation and people wonder why I'm a fearless speaker," my colleague says.

Luie had initiated scientific ballooning at the laboratory in the 1960's with the NASA-funded High-Altitude Particle Physics Experiment (HAPPE, inauspiciously pronounced "happy"). Luie had started HAPPE as a precaution—a way to continue particle physics research—should Congress stop funding the construction of new, larger accelerators.

Some precaution! Two things happened: Congress funded the Fermilab accelerator, and the HAPPE group suffered disaster. At the end of the HAPPE payload's first flight, the fitting that connected the equipment to the parachutes broke and the payload fell freely from a height of ninety thousand feet into the Pacific Ocean. Into the drink went hundreds of thousands of dollars' worth of equipment. Nothing was recovered, and the HAPPE team started a file entitled "Lost at Sea." I inherited this file, and it should have been a warning. Later I would open other files "Lost in the Badlands," "Lost in the Jungle," and "Lost in Space."

When I joined the group our goal was to build a new experiment to discover another putative relic of the big bang. By this time, the big bang was becoming theorists' favored explanation of the origin of the universe, and they could draw on several lines of observational support: for instance, the fact that the universe is expanding; the existence of cosmic background radiation; and the abundance of hydrogen and helium in the universe. Demonstration of the existence of one of the strangest of substances in the cosmos—antimatter—was also considered a clue to a past big bang. It was the search for antimatter that sent my first balloon experiment aloft, my first venture into cosmological terra incognita.

Luis Alvarez, shown in his office at Lawrence Berkeley Laboratory. On his desk are photographs of his heroes and friends: *left to right,* Lord Ernest Rutherford, Walter Alvarez (his father), and physicists Don Gow, Albert Einstein, and Enrico Fermi. LAWRENCE BERKELEY LABORATORY

* * *

Even in the most sober of descriptions, antimatter sounds like the stuff of science fiction. The world we experience—the living and nonliving—is made of matter, namely, protons, neutrons, and electrons that constitute atoms from which everything we know is assembled. What of antimatter? A rock made of antimatter would look like a rock; a person made of antimatter would look like a person; and a star made of antimatter would look like a star. Moreover, material made of antimatter would exhibit the same physical properties as normal matter: antiwater would boil at 100 degrees Celsius and freeze at zero degrees. It is even possible that, like Teller in the poem, each of us could have an antimatter doppelgänger. But if you were to meet your antiyou, forbear to shake hands! For, when matter and antimatter meet, each annihilates the other—mass is completely converted to energy in an incredibly violent explosion. This dramatically demonstrates Einstein's special relativity theory dictum: Mass is equivalent to energy, $E = mc^2$, and vice versa.

We are all familiar with the release of energy when something burns—we experience the energy as heat and light. However, when, for instance, coal or oil burns, a mere one billionth of the mass is converted to energy. In nuclear reactors, which release energy through fission (the breaking apart of atomic nuclei), the conversion is vastly greater, but only reaches 0.1 percent of stored energy. Nuclear fusion (the fusing together of atomic nuclei), which powers our Sun and other stars (also hydrogen bombs) and may one day provide huge energy supplies here on Earth, is more efficient yet, but it still attains only 0.5 percent conversion. In matter/antimatter annihilation, the conversion is 100 percent—in theory, just two pounds of antimatter could yield sufficient energy to power the United States for one day. This inspired some visionaries to muse on the prospects of truly unlimited energy supplies, supposing antimatter could be found and its apocalyptic energy release tamed.

Like matter, antimatter is composed of elementary particles, each of which displays attributes that exactly mirror our familiar matter. Instead of protons, antimatter contains antiprotons, which, among other properties that mirror those of protons, bear a negative rather than a positive charge. Instead of electrons, antimatter contains antielectrons, also known as positrons, which are positively charged rather than negatively charged. Antineutrons, which, like neutrons, carry no electrical charge, display a suite of other physical properties that mirror those of neutrons. For all elementary particles of matter, there are equivalent antiparticles of antimatter.

The English quantum physicist Paul Dirac proposed the existence of antimatter in 1929, as a consequence of trying to merge equations from Einstein's special relativity and from quantum mechanics. Dirac was notorious for his terseness and had little time for lesser minds. During the question-and-answer period following one of his lectures, a listener complained, "Dr. Dirac, I do not understand how you derived this formula at the upper left corner of the blackboard." Dirac replied: "This is a statement and not a question. Next question, please."

Dirac was distressed by his calculations indicating the existence of antimatter—"I thought it rather sick," he lamented—but he did not deny their logical force. (Recall how Einstein's initial reaction to his own theory of general relativity was to deny its ultimate implication—the expanding universe—and to invent the cosmological constant.) Dirac's conclusion was unequivocal: When matter is created from energy, an equal amount of antimatter is also created. For every proton in the universe there should be an antiproton, for every neutron an antineutron, for every electron a positron, and so on.*

Dirac's theory was intriguing, even audacious. His scientific peers could not ignore the notion, and neither did *The New York Times*. In its September 10, 1930, issue, an article described the seemingly bizarre idea under the headline SCIENTISTS ACCLAIM NEW ATOM THEORY. Physicists were eager to put the theory to the test—but how? These days we are familiar with the huge, powerful accelerators—atom smashers—that hurl subatomic particles at each other or at other targets at colossal energies; the resulting collisions (little bangs, so to speak) might generate new particles. The experimenter then has the relatively simple task of noting whether newly generated particles are accompanied by matching antiparticles. In the 1930's, however, particle accelerators were relatively feeble, the largest being the size of a small room.† That was too feeble to generate antimatter.

*Causality requires the existence of an antiparticle for every elementary particle. Causality is the precept that cause precedes effect. It is not obvious that this would be true—as special relativity tells us that moving different observers could see things happen in reverse order—were there not mirror pairs of particles and antiparticles. In a restricted sense, antiparticles moving backward in time behave like particles moving forward in time, and particles moving backward in time are equivalent to antiparticles moving forward in time. This exchange is a space-time mirror symmetry.

†That's a far cry from the superconducting supercollider, or SSC, which is planned to be the biggest accelerator in history. When finished early in the new millennium, it will cover many square miles of Texas prairie. Its purpose: to collide protons and antiprotons, creating intense high-energy conditions that briefly simulate the temperature and density of the big bang at a million millionth of a second (10^{-12} seconds). This will test theories holding that the known forces were part of a single "superforce" in the earliest moments of the universe. The SSC will be a major milestone for the human race as it will represent the halfway point from the present universe to the shortest times and highest energies that we think are possible. The SSC will allow us to observe what happens when two forces—electromagnetism and the weak force—are merged into a simpler, more symmetric force. That will represent achieving a new state of matter that is qualitatively different from all previously observed. It will undoubtedly provide us with new insights on the origin of the universe.

Fortunately, outside anyone's front door there is an immense particle accelerator: the universe. Earth is constantly being bombarded by countless subatomic particles from outer space, some of which have journeyed from distant galaxies, journeys that began before dinosaurs walked the Earth. These streams of particles are called cosmic rays. That's a misnomer, for cosmic rays are charged particles rather than electromagnetic radiation, such as light or radio waves, and are therefore not "rays" strictly speaking. Most cosmic rays are from our own galaxy. Many were probably accelerated to near-light speeds by powerful galactic magnetic fields created by shock waves from exploding stars (supernovae) and other cosmic events. When they enter the Earth's atmosphere, cosmic rays interact with atoms and trigger showers, or "cascades," of other particles, which may, in turn, trigger additional showers. Most cosmic rays are absorbed by the atmosphere, but some of their "daughter" particles make it through the atmospheric gauntlet all the way to Earth. Physicists wondered, therefore, whether they might be able to detect antiparticles among the atomic debris arriving at Earth's surface, thus allowing a test of Dirac's hypothesis.

In 1932, Carl Anderson, of the California Institute of Technology, monitored cosmic rays with a cloud chamber—a device originally perfected in the 1910's by Scottish physicist Charles T. R. Wilson for the purpose of simulating cloud formation. Charged particles leave an ionization trail (atoms with an electron knocked off) in the cloud chamber, giving the supersaturated water vapor seeds around which tiny water droplets can form. For years there was one set up in the front lobby of the main building at LBL. Watching it, one sees a very light mist in the dark chamber, and suddenly a trail of small droplets appears where a charged particle has streaked through. Then the droplets lazily rain down on the bottom of the chamber. Sometimes the tracks point through you!

Anderson put his cloud chamber in a magnetic field, which diverts particles of positive charge in one direction and negatively charged particles in the opposite direction. The radius of curvature of each particle's path is determined by its velocity and mass. Through the window of his chamber, Anderson photographed the tracks of the particles zipping through it. The characteristics of the tracks (the direction they took) effectively served as identity labels of the kinds of particles that made them (positive or negative).

As expected, the detector routinely picked up the passage of protons, electrons, and so on. But on August 2, 1932, Anderson photographed a vapor trail left by a particle that had the same mass as an electron (a negatively charged particle), but whose path bent in a direction taken by positively charged particles. The surprise visitor was the first antimatter particle ever detected—the positron, the antimatter partner of the electron. That positron is presumed to be from a cascade of particles generated by a collision in the upper atmosphere of a high-energy cosmic ray proton or electron with an atmospheric nucleus.

Dirac—who had been justly cautious about his own prediction, and even voiced doubts—was vindicated: He joked that he was smart, but his equation had been smarter. He won a Nobel

Around Edward Lofgren (*center*) are the discoverers of the antiproton—*left to right,* Emilio Segre, Clyde Weigand, Owen Chamberlain, and Thomas Ypsilantis. Behind them is their bevatron. LAWRENCE BERKELEY LABORATORY

Prize in 1933, as did Anderson in 1936. In the 1950's a particle accelerator, called the bevatron, was built at Lawrence Berkeley Laboratory, with sufficient energy to generate its own antimatter. In 1955, Emilio Segrè, Owen Chamberlain, Clyde Weigand, and Thomas Ypsilantis ran an experiment at the bevatron, in which high-energy protons collided with target protons, producing pairs of protons and antiprotons—a quarter of a century after Dirac's prediction. This demonstrated the symmetrical conversion of energy into particles of matter and particles of antimatter. As a result, the existence of antimatter, and the law of symmetry in production and behavior of matter and antimatter, was scientifically established. For their efforts, Segrè and Chamberlain shared the 1959 Nobel Prize in physics.

Like Galileo and Newton before them, twentieth-century physicists were anxious to extrapolate to the heavens what had been learned by experiment on Earth. If little bangs transform energy into equal parts of matter and antimatter, then the same must have been true in the big bang. Half the material universe must, therefore, be composed of antimatter. Where is it? Evidently, there is no significant amount of antimatter on Earth or in the rest of the Solar System. "If it were otherwise," Swedish physicist Hannes Alfvén observed in the 1960's, "the moon-probing rockets would have exploded violently upon impact, and the explosion would have been readily observable on earth." Likewise for the Sun: "If the sun consisted of antimatter it would emit antimatter plasma, or antiplasma, and [this would travel toward Earth where] the aurorae would then glow at a thousand times their present luminosity." (Plasma consists of free electrons and ions, which are atoms with some or all their electrons stripped off. Most matter in the universe exists in the form of plasma, for example in stars.)

Perhaps the antimatter resides elsewhere in the Milky Way? Perhaps in other galaxies? Perhaps even in other universes? (There was even the suggestion that when our universe was formed, a second, identical antimatter universe was also formed; our universe traveled forward in time while the second traveled backward, so never the twain shall meet.) In the 1960's these seemed to be a reasonable range of explanations for why our experience

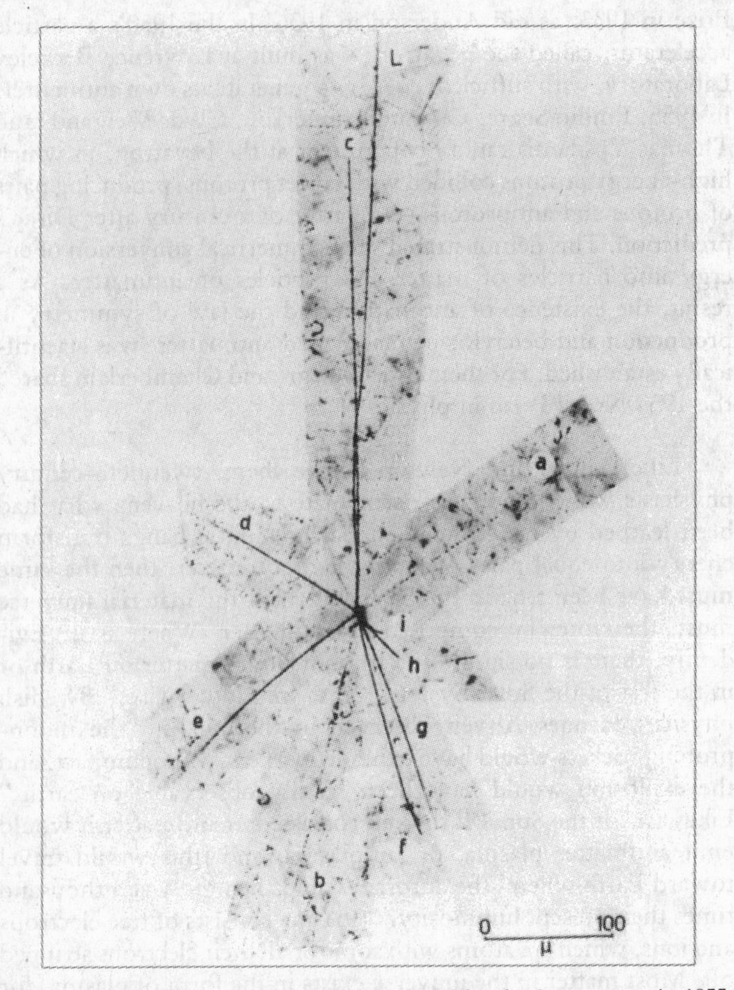

The tracks of the antiproton annihilation found by the Segre group in 1955.
LAWRENCE BERKELEY LABORATORY

is of matter only: Large amounts of antimatter must exist, but (it was argued) we have not yet encountered it directly.

In the 1960's Alfvén proposed an unorthodox explanation for this matter-antimatter asymmetry—for why our experience is of matter only, while, according to theory, large amounts of antimatter must exist somewhere in the universe. Alfvén is an amiable, white-haired Nobel laureate who resembles Santa Claus—without the beard. Born in 1908, a lifelong iconoclast, he compares himself to Galileo (who, he jokes, was "just a victim of peer review"). Alfvén is of special interest to this book because he claims the big bang never happened. Indeed, he is one of a handful of distinguished scientists who still refuse to accept the fiery origins of our cosmos. As an alternative, he offers his antimatter-and-plasma cosmology, which is based on research by the Swedish astrophysicist Oskar Klein.

For much of his life, Alfvén has argued that orthodox astrophysicists overrate gravity's role in sculpting the cosmos. Instead, he says, vast electromagnetic fields and plasma clouds helped mold planetary systems, galaxies, and clusters of galaxies into their present structures. In his 1966 book *Worlds-Antiworlds,* he posited that vast walls of plasma and electromagnetic fields segregate cosmic matter from antimatter, just as the Berlin Wall once divided East and West Berlin. He calls these walls "Leidenfrost layers" after the Leidenfrost effect. You can demonstrate the Leidenfrost effect at home. Get a skillet very hot, then sprinkle a few drops of water on it. You'll see the drops skid around the skillet and bounce off the sides until they evaporate. Leidenfrost explained why this happens: The skillet is so hot that the underside of the water drop is almost instantly boiled into steam. The resulting pressure supports the rest of the drop and insulates it from the heat. So the water drop floats on "air" and "flies" around the pan. At the boundaries between regions of matter and antimatter, the two mix and annihilate each other; the resulting heat and pressure at the boundary would separate the main bodies of matter and antimatter.

Even so, he said, sometimes the Leidenfrost layers are breached: Antiparticles sneak through into the region of ordinary particles, and perhaps find their way to Earth where researchers might detect them. And occasionally, Alfvén speculated, these matter-antimatter encounters trigger titanic explosions that blow enormous amounts of matter and antimatter across the cosmos. That is why the galaxies are receding in all directions—because they are fleeing a matter-antimatter explosion, not because our cosmos underwent a big bang. "The big bang is a myth; it never happened. Our cosmos is eternal, *not* something that sprang into being 15 or so billion years ago," argued Alfvén. "The 'expanding universe' is only an expansion in our part of the universe, and one of countless mini-bangs—each sparked by the mixing of matter and antimatter—that have occurred through eternity."

The most striking result of Alfvén's theory was this: It wasn't inconceivable "that every second star in our vicinity consists of antimatter," he wrote in 1966. "If someone were to claim that Sirius, the brightest fixed star in our firmament, consists of antimatter and not [ordinary matter], we would not have any tenable argument to demolish his claim. Were Sirius to consist of antimatter, it would have exactly the same appearance and emit the very same spectrum as if it consisted of [ordinary matter]. The space around Sirius must then contain antimatter, but this may be separated from the [ordinary matter] in the space around us by a thin Leidenfrost layer, which we are poorly equipped to detect."

Alfvén's theory attracted a lot of attention in the late 1960's, and it begged to be tested. If Alfvén was correct, some of the stream of particles that constantly shower the Earth would be antimatter.

Carl Anderson had detected an antimatter particle—a positron—in his cloud chamber at CalTech, in 1932. That positron almost certainly was generated by cosmic ray interaction with some nucleus in Earth's atmosphere. The observation proved that antimatter exists, but it did not prove that antimatter particles are bombarding Earth as debris from distant agglomerations of antimatter, like those envisioned by Alfvén. Virtually all such antiparticles would be annihilated as they encountered our atmosphere. Hence, to test Alfvén's proposition, it was necessary to loft detection equipment as high as possible, to an altitude where the atmosphere thins to virtually nothing. At that height, an incoming antiparticle is unlikely to hit a nucleus and annihilate in the meager air far above the troposphere. Antimatter detected in the stratosphere is likely to have come from afar, from deep space and not to have been generated from a local collision between two particles of matter. This was our scientific motivation for high-altitude ballooning.

Specifically, my goal was to look for antinuclei that would imply the existence of antistars. True, the detection of an antiproton could mean merely that we had found a rare antiproton produced in a very high-energy cosmic ray collision with material in the near interstellar vacuum or even in the upper atmosphere. We expect to find about three such antiprotons for every ten thousand protons. But, antihelium and heavier antinuclei are produced much more rarely—in parts per trillion and less; were we to detect them, they would be strong candidates for cosmological antimatter. If we found an antihelium nucleus, then that would mean that there was helium left over from the big bang, because essentially all the helium in the universe must have been generated during the first few minutes of this event, as Gamow had figured. Elements *heavier* than helium, however, were produced within stars, as Hoyle and colleagues demonstrated. Therefore, discovery of heavier antinuclei—anticarbon, antinitrogen, antioxygen, and so on—would imply the existence of antistars. Thus began our search for antiworlds.

We launched our balloons from remote sites, such as the U.S. National Scientific Balloon Facility at Palestine, Texas, and sometimes from Aberdeen, South Dakota. Remoteness was an advantage: Balloons sometimes fall to Earth unexpectedly, and no one wanted to "drop" one on an inhabited area. Along with the mental and physical stress of getting the balloon and its payload ready, we were also plagued by wildlife, albeit of the miniature sort. Bugs flew and climbed "into our optics and electronics boxes. Scorpions [came] out of hiding to menace us when we [had] the lights out for our optical calibrations," as Andy Buffington later observed. Everything in Texas is big, including the bugs. These, and the long days of preparation, tried our nerves, and tempers were sometimes short.

To detect antimatter convincingly we had to do what Carl Anderson had done four decades earlier: determine the sign of the charge (positive or negative) of any particle that passed through our detector. We used a liquid-helium-cooled superconducting magnet to deflect the particles. But instead of a cloud chamber, we used a spark chamber. Rather than forming a trail of water droplets along the charged particle ionization trail, the spark chamber uses the free electrons to create a spark—a miniature bolt of lightning. If a high voltage is applied across the spark chamber plates, the ionization track creates a path for the discharge arc of electricity—like the spark of static electricity one experiences on cold, dry days. But we could not keep the voltage on the chamber all the time or it would spark randomly—sometimes missing our cosmic ray quarry. We needed a trigger to know when a particle went through. Fortunately, when an ionized atom recombines with its electron, it gives off a tiny flash of light. Particle physicists had developed special scintillating materials that do this very quickly and efficiently. We used plastic scintillators to detect when a charged particle went through and to measure how much energy it lost by ionizing the atoms in the plastics. Normally, the energy loss is proportional to the square of the particle charge, so that a carbon nucleus—charge 6—gives out thirty-six times as much light as a proton; an oxygen nucleus—charge 8—gives out sixty-four times as much. We set up our equipment so that when a charged particle went through the detector, it automatically applied a high voltage to the spark chamber. The result: flashing lights that signaled the event number and the amount of light emitted in the scintillators. Our plan was to analyze the filmed events for evidence of a particle with incongruous properties—for example, one with the mass of a carbon nucleus but a negative charge.

The flash of the spark and the lights were recorded on film. We also installed (on a later flight) a microphone to transmit the sounds of the firing spark chambers and the whirring of the film as it advanced through the camera. These sounds gave us some comfort from hearing the equipment was functioning correctly. Not so comforting was the eerie background creaking and popping of our fiberglass-shell gondola. The sounds were caused by changing pressure differences between the interior and exterior of the gondola during the ascent.

Our instrument had advantages and disadvantages over its predecessors in the antimatter-detection business. On the one hand, the superconducting magnet could detect particles some fifty times more energetic than could an older method that employed stacks of emulsion, a form of photographic film. The more energetic a particle, the farther it can travel from its origins; hence we were sampling cosmic rays from far into our galaxy and beyond. On the other hand, the magnet had to be kept cold during flight in a liquid-helium cryostat, which was not only hazardous but also made the payload heavy.

The flight progressed well. After ten hours it was time to release the gondola. From the experience of others, we knew this was one of the times when things were most likely to go wrong—and in a big way. Witness HAPPE's first inauspicious outing. We were luckier, however. On signal, the gondola released and its parachute opened, bringing its payload safely to the ground—more or less. During its descent, the gondola was swinging alarmingly under its chute. We had carefully built a polystyrene foam base beneath the gondola, to cushion its impact, but its sides were unprotected. As luck would have it, the gondola landed among trees, and as it neared the ground its swinging motion brought it into violent contact with a hefty trunk, crushing one side and destroying part of the equipment.

When Andy Buffington and I reached the swampy pine forest landing site, we were aghast at the damage to the gondola, and dashed to see if the cameras were intact. If the film had been exposed to light, our data would have been wiped out. Fortunately, the cameras had survived the impact, but we looked despondently at the crushed gondola shell, imagining the worst. Fortunately, only a couple of mirrors were broken.

During its ten hours aloft the detector had recorded fifty thousand events, itinerant particles leaving their imprints on hundreds of feet of film. We rigged up an automatic device to scan the film and discount all normal patterns, but that still left a handful of tracks that seemed to curve the "wrong" way, a possible signature of antimatter. Was this our quarry? Luie was excited by our results and told us to prepare a dossier on each of the anomalous tracks, describing their properties in detail. Armed with our reports, Luie personally studied each antimatter candidate and managed to explain all of them—except one. We called it cosmic ray event No. 26262, named for the number emblazoned by our lights onto each picture.

In the top and middle plastic scintillators, No. 26262 had

produced as much light as an oxygen nucleus, but it had released an unusual amount of energy in the bottom scintillator. A shiver ran down our spines when Luie mused: "Could the excess energy have resulted from an antioxygen nucleus annihilating with an ordinary atom?" We were extremely nervous that our news might leak out prematurely. Imagine the embarrassment if, following premature publicity, No. 26262 proved to have an ordinary, non-antimatter explanation. Luie assiduously pushed us into further and further testing, and he'd frequently say, "Let's be sure to give this [possible antioxygen] event a decent burial." By this he meant we should be certain of our explanation and not just either dredge up some excuse to get rid of it or trumpet it as a great discovery without convincing evidence. Luie was a stickler for precision. He knew how easily instruments and people can be misled. He hadn't forgotten how, in the early 1950's, he and a colleague had detected a curious laboratory effect that looked like a slow form of nuclear fusion. The news media had heralded the story and speculated about a potentially boundless supply of energy. Soon Luie realized it was only an odd chemical effect, muon catalysis, and not a useful energy source after all.* He was determined to avoid such needless brouhahas in the future.

As the weeks passed, No. 26262 continued to defy our attempts to assign it to ordinary matter, and we became increasingly convinced that we had detected the first cosmic antinuclei. We speculated about the possible existence of antistars and antiplanets, and joked that as soon as we announced it, the Vatican would get to work on the antipope question.

However, even though we could not prove that No. 26262 was a particle of matter, neither could we prove it to be an antiparticle. An extremely energetic, regular oxygen nucleus would have a very straight path through the magnetic field. But, small errors in the viewing mirrors' optical alignment and our reconstruction of its track could have made the path appear to curve the wrong way. If its collision in the bottom plastic scintillator happened to have been a hard one, it would have deposited much of its high kinetic energy there. Factoring in these other

*The "cold fusion" furor of 1989 was another example of scientists who thought they had detected a similar effect.

possibilities, we calculated a three-to-one chance that No. 26262 was an antimatter particle. Three to one might sound like good odds, but physicists are more demanding than that for an important new finding. The detection of an extragalactic antimatter nucleus would have been such an extraordinary discovery, one portending so many important implications, that our colleagues (all natural skeptics, like Luie) wouldn't have accepted it without equally extraordinary evidence. Our colleagues told us they would not believe No. 26262 was antimatter unless the odds in its favor were at least a million to one.

We decided it would be prudent to write off No. 26262 as a fluke, and published a paper in *Nature* saying that the balloon flight had found no evidence of antimatter. Meanwhile, we planned an improved follow-up experiment to find out if No. 26262 might be evidence of antimatter. We reduced the amount of material in our apparatus to minimize the number of occasional interactions between the particles and the detectors that might produce illusory "antimatter" signals. We simplified the optics by having the camera directly view the curved particle path. This meant the balloon's gondola had to be longer and more egg-shaped, with the magnet and cosmic ray track detectors on one end and the cameras and electronics on the other end. We also built a more powerful magnet. The more powerful it was, the greater the degree to which it could curve the path of an incoming particle, and the greater the curve, the greater our chance of correctly identifying the sign of the particle's charge. If there really was antimatter out there, we were going to find it.

During the following years we launched six more balloons, and some of them under distressing circumstances. On the evening of May 28, 1977, what turned out to be our last payload and balloon lifted off from an airport on the outskirts of Aberdeen, South Dakota. All around stretched the empty prairie—off in one direction lay the Badlands of South Dakota. Far above, our balloon drifted lazily through the night—our equipment gathering data. Charles Orth sat in our trailer monitoring telemetry from the balloon instruments and sounds from within the gondola picked up by an on-board microphone. Everything was going fine. Suddenly at 7:25 A.M. he jumped up and screamed: "It's falling!" Everyone stared at him. He waved at the instruments.

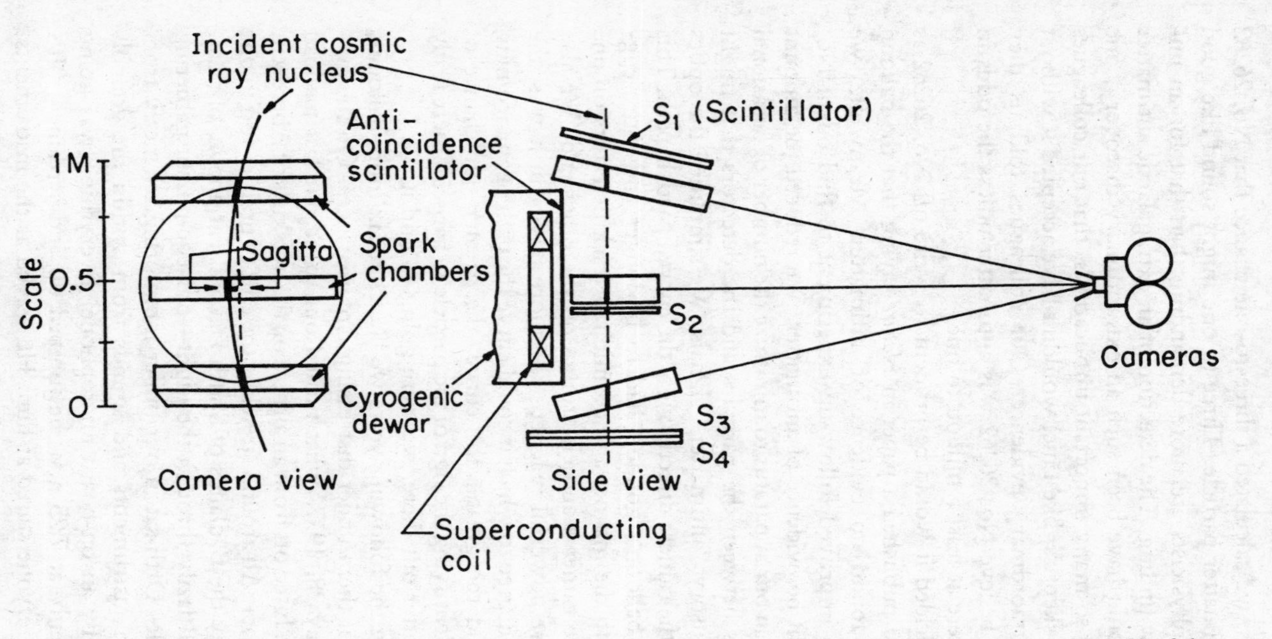

A sketch of the new payload showing a camera direct view of cosmic ray nuclei bending in the magnetic field of the superconducting magnet. The amount of bending (labeled "Sagitta") shows the power of measurement, and the direction of the sagitta separates matter from antimatter.

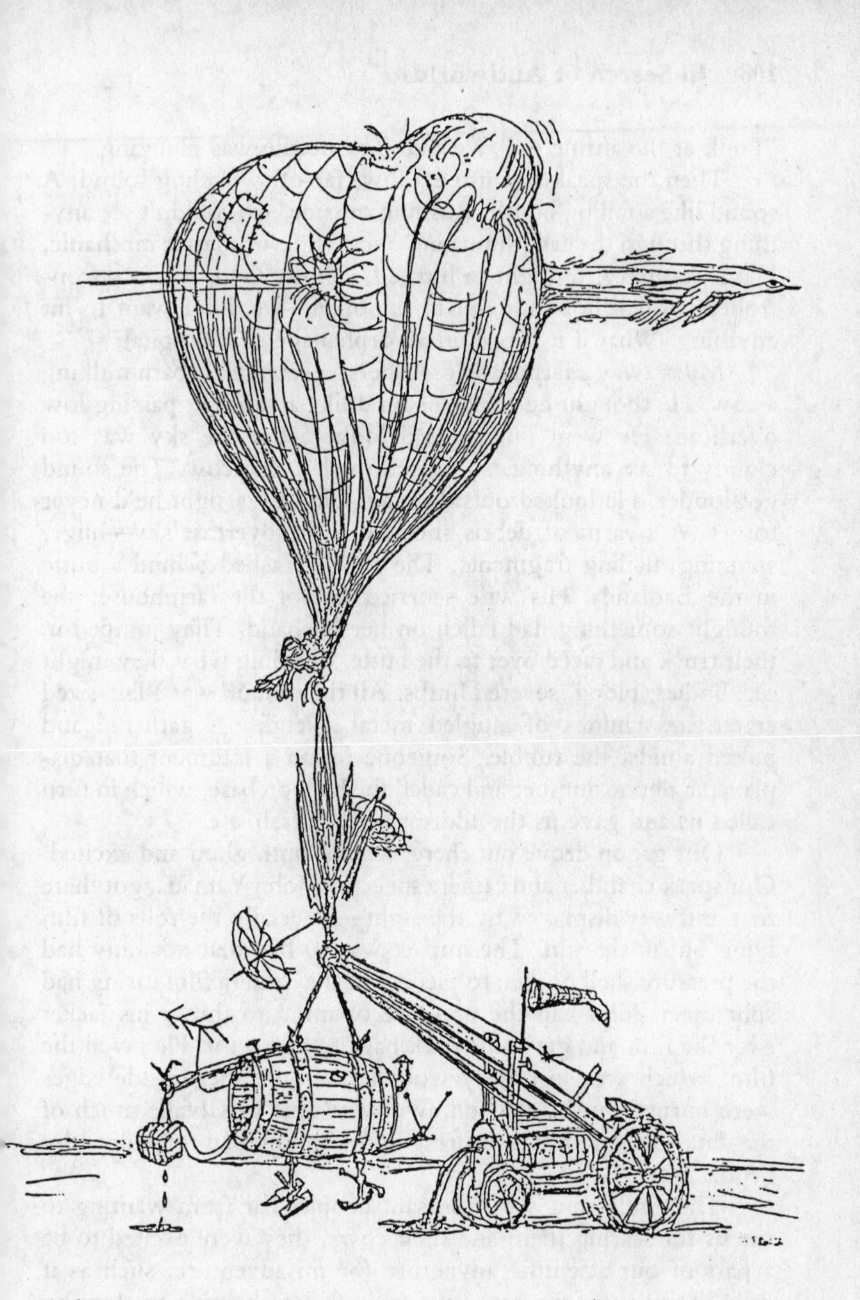

Hal Dougherty's spoof sketch of ballooning in South Dakota farm country.

"Look at the altimeter!" We did. The needle was plunging.

Then the speaker emitted a low, far-off whistling sound. A sound like a falling bomb. Rushing outside, we couldn't see anything through the early-morning overcast. Our master mechanic, Hal Dougherty, thought to himself: Oh my God, it's an uncontrolled drop. I hope it's out in the open—we don't want to hit anything. What if it drops on an orphanage or a hospital?

Miles away a farmer, Mr. Anderson, was in his barn milking a cow. He thought he heard a sound like an airliner passing low overhead. He went outside and looked, but the sky was too cloudy to see anything, so he returned to the cow. The sound got louder. He looked outside again and saw a sight he'd never forget. A swarm of debris shot from the overcast sky—huge, spinning, flailing fragments. The debris crashed behind a butte in the Badlands. His wife scurried out of the farmhouse; she thought something had fallen on her husband. They jumped in their truck and raced over to the butte, dreading what they might see: bodies, blood, severed limbs. All they found was a fair-sized crater and chunks of tangled metal. Neighbors gathered and poked amidst the rubble. Someone found a fragment that displayed a phone number and called the balloon base, which in turn called us and gave us the address of the crash site.

Our group drove out there, feeling both glum and excited. Our spark chamber and camera specialist, John Yamada, got there first and was dismayed by the sight—especially the rolls of film lying out in the sun. The impact was so bad that not only had the pressure shell broken to pieces but the camera film casing had split open. John had the presence of mind to throw his jacket over the film and get some dark bags to protect it. He saved the film, which was so tightly wound that only the outside edges were burned out by the sun. We were able to salvage much of the data—though the measuring and scanning had to be done by hand.

The Andersons were pleasant people. Far from wanting to sue us for scaring them and their cows, they were excited to be a part of our scientific adventure (or misadventure, such as it was). They spent the next two or three days helping to clean up the mess. Back in Berkeley I opened a new file folder: "Lost in the Badlands."

During all the years leading up to this last flight we found not a single convincing sign of cosmic antimatter. My infatuation with ballooning was over.

Frustrating though they may be for the researchers who collect them, negative scientific results are as important as positive results. True, negative results cannot be conclusive. But they define the boundaries of our knowledge; they tell us what we can reasonably believe and what remains uncertain, unresolved—in short, what remains to be studied. We published our negative findings (in *Nature* and the *Astrophysical Journal,* and in 1975 the surviving members of the team—Smoot, Buffington, and Orth—reported in the *Physical Review Letters*). The American Institute of Physics recognized one of our balloon flights as one of the world's twelve outstanding physics experiments of 1973.

To this day our work remains the most stringent limit published on complex antimatter nuclei in cosmic rays. We said that antimatter nuclei, such as carbon and oxygen, assuming they exist, are rarer than one in ten thousand in our part of the universe. Conceivably they could exist at levels of one in a million or a billion or a trillion, or they might not exist at all; in any case, they're extremely rare.

Our result contradicted Alfvén's hypothesis of the history of the universe: If his antimatter-and-plasma cosmology was valid, then we should have seen at least one antimatter particle for every ten thousand detected. We didn't.*

But what of the big bang theory? Is that not falsified too? If evidence from little bangs can be extrapolated to the big bang, then there should be equal amounts of antimatter and matter in the universe, and yet we found no evidence of it. Perhaps such an extrapolation is inappropriate. Perhaps there has always been an excess of matter over antimatter.

In 1967 the well-known Russian physicist Andrei Sakharov proposed the core of a solution to the puzzling matter-antimatter

*For the record, Alfvén disagrees. He believes the particle energies measured by balloon experiments are too low to detect antimatter energetic enough to penetrate the Leidenfrost layer.

asymmetry problem. His hypothesis involved unusual conditions in the early universe—conditions outside the experience of experimental particle physics—and so went unnoticed for almost a decade. Not until new theories emerged in the mid- to late 1970's was Sakharov's proposal recognized.

A courageous, principled man, Sakharov led a life that was the epitome of irony. As a young man, he developed the Soviet hydrogen bomb for Stalin; as an old man, he became a political dissident, with his equally outspoken wife, Yelena Bonner, by his side. The Soviet authorities condemned him to years of internal exile. Even so he managed to continue scientific research. In the late 1980's he was freed by Mikhail Gorbachev and lived long enough to be elected to the Soviet legislature; he died in 1990, shortly before the collapse of the dictatorship he had once served and, later, assailed.

Sakharov's proposal, as viewed in the light of modern cosmology, is as follows: Processes in the first instant (within the first one millionth of a second) after the big bang produced a slight excess of matter over antimatter; matter and antimatter particles then self-annihilated in a cataclysmic release of energy, leaving the slight excess of matter, which went on to form the known universe. The huge release of energy in the matter-antimatter annihilation was manifested as cosmic background radiation photons (packets of light energy), which vastly outnumber baryons (the collective term for heavy particles such as protons and neutrons) by some one billion to one. In this scenario the present universe consists almost exclusively of matter (perhaps with vanishingly small pockets of antimatter that escaped the early orgy of annihilation) and the billion-to-one ratio of cosmic photons to baryons, which is observed today.

We know that something of this sort must have occurred in the first instant of the universe, with a slight excess of matter over antimatter. If, instead, matter and antimatter had been produced in exactly equal abundance at the moment of creation, the universe would be a very different place. A vast annihilation event would have occurred, leaving only very few particles of matter and antimatter in scattered, isolated remnants, and an all-enveloping sea of photons. The night sky would have only the soft hum of microwave radiation, without its glimmering pattern of galaxies and stars.

Sakharov's proposal required that two "givens" of particle physics should be violated: (1) the conservation of baryon number, and (2) CP symmetry. Although these sound arcane, they can be described simply. The first merely restates the fact that energy is always converted into equal amounts of matter and antimatter—that is, there can be no net production or net loss of baryons (matter) over antibaryons (antimatter) in any reaction. The second states that in their chemical and physical reactions, matter and antimatter are identical. In 1956 Val Fitch and Jim Cronin had discovered that the strange kaons violated CP symmetry at a level of one part in a billion. So nature has a "toe in the door" toward CP symmetry violation. As a result, matter can behave slightly differently than antimatter. If baryon number is violated, then there is the possibility of producing a small excess of matter over antimatter.

During the late 1960's and early 1970's, several theoretical and experimental developments occurred that encouraged the idea that the once-forbidden baryon number violations could in fact occur. Most important of these was the formulation of a theory—strictly speaking a family of theories—that sought to unify three fundamental forces of nature, the weak, strong, and electromagnetic forces. They are known as grand unified theories, or GUTs for short, and were pioneered by Pakistani physicist Abdus Salam, University of Texas physicist Steven Weinberg, and Harvard physicist Sheldon Glashow.

The weak, strong, and electromagnetic forces behave differently in the universe as we experience it. But long ago (according to GUTs), at the unimaginable temperature of the first instant of the universe (within 10^{-34} seconds, or one ten-millionth of a trillionth of a trillionth of a second after the big bang), they were essentially the same and operated in the same way on particles. According to GUT theorists, baryon number will not be precisely conserved. This may sound like the proverbial rabbit out of the hat, but it is in fact a central tenet of all GUTs.

My optimistic expectation is that baryogenesis—the production of the excess of matter over antimatter—occurs at the time when the electromagnetic and weak forces unify. According to the Weinberg-Salam electroweak theory, this happens at energies equivalent to about 100 proton (or neutron) masses. The superconducting supercollider will reach that energy level, so we may learn the answer in about ten years—about the year 2005.

Producing an excess of matter is, however, only a start. In physical systems at equilibrium, reactions flow readily in both directions: The excess of matter might disappear just as fast as it appears. However, the big bang universe is expanding and cooling. Once the temperature has fallen to a level too low to drive the matter-producing reaction in the reverse direction, the small excess of baryons produced becomes a permanent component of the cosmos.

It may appear from this description that the universe—and our existence in it—was the result of a lucky break: a slight excess of matter produced as a result of violating rules at the right time. This may be but one of countless possible outcomes of that minuscule slice of time following the big bang, or it may be an inevitability given the laws of physics that operated then. We do not yet know. This pursuit of antimatter, which has obsessed many of us for so long, has forced us to face one of the key properties—and toughest challenges—of cosmology: namely, the need to extrapolate physical laws discovered on Earth to the cosmos. This extrapolation seems to work admirably for most of the history of the universe. But it may begin to break down as we approach the instant of the big bang. That, indeed, is terra incognita.

Chapter 6

Spy in the Sky

Well before our final, ill-fated balloon mission crashed in the Badlands of South Dakota that day in May of 1977, I had turned some of my thoughts away from the search for antimatter, prompted by Luie Alvarez, our team leader. In early 1974, sensing that we had probably gone about as far as we could with our existing instrument, he had given me and my colleagues some good advice. He said in essence: "Before you rush into continuing with a new, improved version of the experiment you've just done, sit down and take stock. Think deeply about what the important problems in physics are. And think particularly deeply about the important opportunities in physics, opportunities that have emerged thanks to new technologies and recent experimental results. Take a month or two to think about it; study what the most interesting and important problems are. Then decide what you want to do next." It was the kind of good advice that flows from wisdom and experience.

I was approaching thirty and had little time to waste. None of us had. We—that is Andy Buffington, Terry Mast, Rich Muller, Charles Orth, and I—took Luie's words to heart and started assessing possible future projects. After weeks of reflection and freewheeling discussion we each moved toward our own decisions. Andy and Charles opted to continue the antimatter and cosmic ray experiments. (I continued to collaborate with them on this until the Badlands incident.) Rich became interested in the development of accelerator mass spectrometry to measure radioactive species. Later he developed an automated telescope system to detect supernovae that, he hoped, would provide a better measurement of the Hubble constant and the deceleration of cosmic expansion. Terry considered and dabbled in many projects before settling to work with Jerry Nelson on what is now the Keck ten-meter telescope, the largest telescope in the world, atop Hawaii's Mauna Kea.

One place we found inspiration was in James (P.J.E.) Peebles's 1971 book, *Physical Cosmology,* which to my generation of observational cosmologists is what Darwin's *On the Origin of Species* was to Victorian naturalists—a blueprint for future research.*

One section of Peebles's book was titled "Applications of the Primeval Fireball." The first application of the "possibly discovered Primeval Fireball"—an allusion to the cosmic background radiation seen by Penzias and Wilson—was "The Aether Drift Experiment." Peebles's cogently argued discussion of this topic, with its scholarly references, made me realize my future was in exploring the cosmic background radiation. The cosmic background radiation is a relic of the big bang, a low-level microwave hum (a bath of light at just 3 degrees Kelvin) that pervades the cosmos. Most astrophysicists considered this distant echo of creation to be pregnant with clues to the subsequent evolution of the universe, if scientific ingenuity could ferret them out. Peebles's reference for the original idea was a 1967 paper by Dennis Sciama of Cambridge University. Sciama argued that the cosmic background radiation could be used as a tool to check whether the local inertial frame was rotating against the distant matter in the universe. A successful measurement would produce a five-thousand-fold increase in accuracy testing Mach's principle. I was enthralled: One could use the cosmic background radiation to learn about the rotation of the universe.

In 1949, the famed mathematician Kurt Gödel had found a solution to Einstein's general relativity equations in which the

*Steven Weinberg's 1972 book, *Gravitation and Cosmology,* also inspired a number of young particle theorists to enter cosmology.

entire universe is rotating, an audacious hypothesis that also implied that time travel is possible. Later, there were other models, both more conventional and more bizarre than Gödel's, of rotating and distorting universes, but not all involved the tantalizing prospects of time travel. In all of them, however, an effect of rotation was to cause characteristic temperature variations across the sky. If these existed, we should be able to detect them. One could also determine if the universe is expanding symmetrically, as the Hubble law suggests, or whether it's expanding faster in some directions than others. If the latter proved to be the case, it would mean that the universe—and the conditions that produced it—were more complex than current theory allowed.

The most worrisome thing (to theorists) about the big bang is it required space-time to begin at a single mathematical point, a phenomenon known as a singularity. Strange as it seems, this point would have zero size and infinite density. At such a point, the laws of physics break down and the universe is effectively beyond mathematical description. Theorists balked at such a prospect because, like Einstein, they did not want the laws of physics to break down and have chaos reign. This embarrassing prospect might be avoided, it was suggested, if the universe were expanding asymmetrically. The universe could then oscillate, alternately expanding and contracting, because the "big crunch" at the end of the contracting phase could be avoided. If one direction was collapsing more quickly than the others, then the universe would pancake, pass through, and start expanding outward again before the other directions arrived.

This was unexplored but important territory, so I knew that if I were to test it experimentally, I would be certain to find something interesting.

The title of Peebles's commentary—"The Aether Drift Experiment"—was an allusion to an idea that has permeated cosmological perspectives since early Grecian times. Originally conceived of as a pure, unchanging, "ideal" fluid through which the heavenly bodies moved, the aether later (under Aristotle) came to be thought of as an exquisite crystalline substance. Crystalline spheres of aether encircled the Earth, carrying the Sun, Moon, and stars with them in perfect orbits as they revolved. Newtonian cosmology abolished the need for aether, as space was said to be an empty void. But the notion returned a century later and continued into Victorian cosmology, this time as an ethereal substance through which light waves propagated. Scientists diligently searched for evidence of this luminiferous aether, but all such efforts were ultimately declared fruitless by a famous experiment performed by Albert Michelson and Edward Morley in 1887.

The two American physicists reasoned as follows: As Earth orbits the Sun, it moves through the aether. Hence one direction is "upstream" and the opposite "downstream," in terms of aether flow. Consequently, a beam of light pointed upstream would move more slowly through the lab, and one pointed downstream would travel more quickly. Any clear sign of a variation in *c*, the speed of light, as the Earth rotated would prove that the aether existed. In an utterly unexpected result, they found no variation in *c* at all. The aether did not exist.

The loss of the aether deprived science of Newton's absolute space—a "universal frame of reference." According to this cherished concept, there is a framework and solidity to space and time. A distance of a meter is always a meter, never a millimeter shorter or longer; a second is always a second, never passing more slowly or quickly. True, there is no way for a human to say that, for example, a ball is *absolutely* moving at a certain velocity. All one can say is that it has moved a certain distance over a certain period of time *relative* to a specific spatial frame of reference—like the surface of the Earth. The "true" motion of the ball is impossible to know, for while the ball rolls along at, say, one foot per second, other motions are occurring on a grander scale: The Earth is rotating and is orbiting the Sun, the Sun and its entourage of planets are moving through the Galaxy, the Galaxy is rotating, and so on.

What, then, is the "true" motion of the ball? One can say that its true motion is the sum of all the motions described earlier. Yet, how can one determine that "true" motion? What is the ultimate frame of reference against which the ball's motion is to be judged? In the eighteenth century, Newton assumed this frame of reference was something he called absolute space, which he seems to have equated with God. Later, the notion of space was embodied in the aether, which was thought to fill the entire universe, but which Michelson and Morley demonstrated did not exist at all. The absence of aether was no surprise to Einstein, however; his 1905 theory of special relativity had no need of it. All observers, Einstein argued, regardless of their frame of reference, perceive light propagating at exactly the same speed—186,282 miles per second, never faster or slower. This fact is the foundation of special relativity, according to which there is no absolute space or time, no ultimate frame of reference against which changes in velocity can be judged.

Yet not all is lost, Peebles said in his *Physical Cosmology*. The cosmic background radiation shares some of the qualities that the Victorians had erroneously ascribed to the aether. The cosmic background radiation pervades all space, he said, and hence could be used as a surrogate for the mythical absolute space envisioned by Newton and rejected by Einstein, but *without* violating special relativity. The cosmic background radiation would be a universal frame of reference against which absolute motion may be detected. This, then, would be my toehold for exploring the universe.

The cosmic background radiation is virtually homogeneous in all directions. Such a homogeneity may not seem a promising frame of reference for detecting motion in the universe, as it possesses no identifiable landmarks—it is like being in a thick fog. However, by virtue of the Doppler effect it is possible to detect motion through it. If the Earth is motionless with respect to the rest of the universe, then the cosmic background radiation is uniform in all directions, with exactly the same equivalent temperature in all directions. In other words, the radiation would be said to be isotropic. If the Earth is moving, however, there is a smooth variation in temperature across the sky, because of the Doppler effect. In the direction in which the Earth is moving, the cosmic background looks warmer; in the direction of recession, cooler. (It is like driving through a vertically falling rain. The front window gets more rain hitting it harder than the back one.) The Doppler effect produces a dipole anisotropy (two poles: a warm pole and a cool pole).

The amount of warming and cooling is proportional to the speed of motion compared to the speed of light, and the direction

The U-2 aircraft from NASA-Ames flying above the Golden Gate Bridge; fog and San Francisco in the background. The high-altitude U-2 jets were famous for reconnaissance during the Cold War. Now they are used for scientific research. NASA-AMES

of the dipole lines up with the direction of motion. The Sun and Earth orbit the galaxy at about 250 kilometers per second. Sciama and Peebles pointed out that this meant there would be a dipole anisotropy at the level of 0.08 percent—nearly one part in a thousand.

I realized that the technical and logistical challenges of the aether drift project were likely to be enormous. Having Luie's support was therefore going to be crucial to the project's success. A decade after winning the Nobel Prize, he carried a lot of weight in scientific and political circles. He was properly skeptical about our venture, and even vehemently told Rich Muller that we would never see a dipole. Rich concluded Luie was overreacting to the Michelson-Morley experiment of 1887. Their failure to find the aether had been so dramatic that, ever since, physicists had felt an instinctive prejudice against any effort to detect a universal frame of reference; Luie shared that prejudice. Still, with some effort we managed to talk Luie into backing the experiment. We then changed the name, calling it "the *new* aether drift experiment," before seeking funding.* Bob Birge, then head of LBL's Physics Division, provided seed money to start the project.

Luie's skepticism was understandable in one sense: The dipole would be difficult to detect. The new aether drift signal would be extremely faint, probably varying by one part in a thousand of 3 degrees Kelvin (three thousandths of a degree), which, at that time, was at the limit of what our instruments could sense. We would have to distinguish the faint temperature difference without confusing it with a myriad of "noise," such as that from celestial objects—stars, interstellar dust, and galaxies. Since the big bang, a great deal has happened in terms of cosmic evolution, and it all generates noise. In addition, there would be thermal radiation from Earth and from the instrumentation. Also, Earth's atmosphere is a source of microwave noise, because microwave radiation is emitted by oxygen atoms and water vapor. The instrument would therefore have to be placed at a high altitude—above much of the atmosphere—where it could monitor not only the cosmic background radiation but also the Galaxy and atmosphere. Once we knew how loud these sources of noise were during our anisotropy measurements, we could subtract them from the overall radiation measurement. Any residue might be the cosmic background radiation signal. In short, "listening" for cosmic anisotropy would be like listening for a whisper during a noisy beach party while radios blare, waves crash, people yell, dogs bark, and dune buggies roar.

We faced at least two big questions. First, how could we make an instrument that was sensitive enough to detect anisotropy, but design the measurement so that it would not be swamped by radiation from the sky, atmosphere, and instrument itself? Second, upon what kind of "platform" should the instrument make its observations? A mountain? A balloon? Or something else? We knew that Brian Corey and David Wilkinson, of Princeton University, were also planning to search for the dipole,

*Sometimes even the "new" was not enough. When my girlfriend, Constantina Economou, visited my lab and saw the sign, she was amused and asked, "Haven't you heard of Michelson and Morley's experiment? We studied it in philosophy class."

almost certainly using a balloon system. Dirk Muehlner and Rainer Weiss of MIT were also planning a balloon experiment. Adrian Webster of Cambridge was testing the feasability of mountaintop observations, as was Paul Boynton of the University of Washington. We wondered who would be first to find anisotropy, but the technical problems were so keen that we couldn't afford to make a mistake through being hurried by the specter of competition. In any case, I'm something of a perfectionist—some would say obsessionist—and I've always put faith in the power of producing the very best that is possible.

While mulling the questions of what kind of platform to use, and mindful of the competition, we began recruiting a larger team for the dipole project. An important catch was Marc Gorenstein. I had known Marc while I was a graduate student at MIT. He grew up in Boston and had always been scientifically inclined, with a touch of the romantic—the kind of kid who, at age ten, was "astounded" when he realized people were made of atoms. Besides Marc, Rich, and me, the new aether drift team included Jon Aymon, who handled much of the computer programming. He did the programming needed to transfer data from the magnetic cassettes to our "number-crunching" computer. He and I wrote most of the data-processing and early analysis software. Another invaluable ally was our amiable master craftsman, Hal Dougherty, a veteran of many Lawrence Berkeley Laboratory projects. He built almost everything mechanical on the project, plus a model of the equipment used to check the detector's mechanical design and rotation system. Then there was John Gibson, who made the electronic systems and housekeeping devices such as temperature sensors. Robbie Smits designed and built the experiment's very fast, very stable rotation system.

Another crucial team player was Tony Tyson, who was visiting Lawrence Berkeley Laboratory while on six months' leave from Bell Labs. Tony was also my roommate during his visit. He was an expert on sensitive instrumentation, thanks to his experience with gravity-wave detectors. He knew ways to prevent our measurements from being hampered by vibrations from the observation platform; our work with balloons, which are relatively noise-free, hadn't prepared us for the latter problem.

* * *

For almost a decade, astronomers had been studying the cosmic background radiation with a variety of instruments, such as radio receivers and bolometers, which were sensitive to the extremely faint microwave radiation from the sky. We decided to use a type of radio receiver known as a differential microwave radiometer, or DMR, the ancestor of which had been invented in the 1940's by Princeton's Robert Dicke (leader of the scientists who pointed Penzias and Wilson toward the correct meaning of their discovery of the cosmic background radiation in 1964).

The DMR didn't measure the absolute temperature of a given point in the sky; rather, it measured the difference in temperature between two points in the sky. Hence the term "differential." Whereas a one-antenna radiometer would say, in effect, "The temperature at point A is 2.725 degrees K," a dual-antenna differential radiometer would say, "The temperature difference between point A and point B is 0.002 degrees." Since in those days good receivers had noise of their own that was equivalent to 400 degrees, an *absolute* radiometer would have to be stable to better than one part in a hundred thousand. But if we could match the antennas well, then the DMR had to be no better than about one part in a thousand. So we designed a symmetrical DMR, which rested on a turntable that could rapidly and repeatedly spin 180 degrees back and forth. That way, each antenna view was interchanged, and the signal from the sky should change sign while any false signal from any DMR asymmetry would stay constant. This double check of each measurement vastly increased the instrument's overall sensitivity and trustworthiness.

We knew that building such an instrument would be a technically demanding task. Just how difficult would depend, however, on where we put it. Our goal was to get as high above Earth's surface as possible—as far as we could above military radar, atmospheric oxygen, water vapor, and other sources of microwave interference. At about this time I had begun fantasizing about putting our instruments on Earth-orbiting satellites, and was already making plans to explore possibilities with NASA. Such a prospect was a long way in the future, however, and meanwhile I had to be content with more mundane solutions.

One possibility was atop White Mountain in eastern California, site of a proposed University of California observatory. It was a cold, dry place, 14,000 feet in altitude, where scientists panted for air and altitude sickness was common—a scenic spot for a vacation but a grueling place to work. The one advantage of such a location was that we could build the differential microwave radiometer just exactly as we wanted it—the observatory could accommodate anything of reasonable size and shape. Moreover, repairs would be relatively straightforward. We also talked about launching the instrument on a balloon, which would give us greater altitude, as high as 120,000 feet. I was less than enthusiastic about the idea, given my recent experiences—if possible, I never wanted to balloon again.

Then, in the fall of 1973, a fortuitous event occurred. I heard a lecture by Charles Townes, a Berkeley Nobel laureate famous for developing lasers. Townes had also played a key role in developing the Kuiper Airborne Observatory (named for famed astronomer Gerard Kuiper), a C-141 military transport that was then being converted for use as a flying observatory. I thought to myself, Why don't we put the DMR on a plane? It would rival the altitude of a balloon, and still give us more hands-on control over the instrument. Rich was initially very skeptical at the idea, but Luie was intrigued. He liked airplanes and knew a lot about them; he had won the Collier Trophy in 1946 for developing instrument landing during World War II, thereby saving the lives of many Allied pilots. Luie happened to know Hans Mark, then the director of NASA-Ames, and this provided us with a perfect entrée into the bureaucratic process of having a proposal approved. Luie immediately wrote Hans a letter explaining our needs.

I visited the NASA-Ames facility, located in Mountain View about thirty miles south of San Francisco, to check out the Kuiper Airborne Observatory, and quickly encountered two problems. First, the C-141 had only one window for astronomical observations, while the DMR would have two horn antennas pointing in different directions. That wouldn't work for us. A further objection was that the plane didn't get high enough to eliminate enough atmospheric noise. It seemed we might have to resort to balloons after all. Then Luie and I cooked up an alternative—the U-2 spy plane.

First built in the mid-1950's, in the early years of the Cold

War and before the advent of spy satellites, the U-2 reconnaissance plane—or Dark Lady, to its intrepid pilots—was America's best hope for monitoring Soviet forces and weapons development. U-2s were essentially gliders with powerful engines. Extremely lightweight, they flew high in the stratosphere, seventy thousand feet or higher, and covertly violated Soviet airspace to photograph activities far below. They flew so high, in fact, that the pilots wore "space suits," the ancestors of those worn several years later by astronauts. At first the Russians didn't publicly protest the flights but did so privately through diplomatic channels—and they were willing to shoot down any aerial intruder. Their antiaircraft rockets were slow and inaccurate, but they were getting faster and more precise. To make the U-2 harder to spot, its developers painted its underside dark blue to blend with the upper atmosphere. Even so, numerous pilots were shot down. When I was a kid, the world was stunned to hear that the Soviets had shot down Gary Powers. It proved to be a major diplomatic incident, and the U-2 was a secret no more.

When Luie and I first talked about the U-2, we saw it would make an ideal platform for our instrument. The aircraft flew extremely high and was very stable, just right for our purposes. We went to Ames to talk about flying the experiment. Luie didn't pressure Hans Mark—an imposingly tall guy with a crew cut and German accent who looked like a general—into offering the spy plane. We waited until he said what we hoped to hear: "You ought to try the U-2, it's perfect."

We were thrilled. We would launch our DMR aboard a spy plane. It was hard to believe. I recall the eerie pleasure of getting a copy of the *U-2 Users' Manual*. Hans Mark's enthusiastic support gave us confidence we could see the new aether drift experiment through the NASA bureaucracy. "It certainly would be fun for all of us to work together with your group on this," he wrote Luie in a letter dated October 8, 1974. We were on our way.

Despite initial cost uncertainties, we were delighted with the U-2 engineering team at Lockheed. They had built the spy plane in the 1950's and knew it intimately. Marc Gorenstein and I traveled to the Burbank airport, home of Lockheed's "skunk-works," which was the birthplace of some highly secretive military technology (for examples, the U-2 and the SR-71 Blackbird).

The Lockheed experts were excited to help us out. They said photo reconnaissance technology was becoming pretty old hat, so they were eager to conquer new fields.

Our first major challenge was to merge the differential microwave radiometer with the U-2. We couldn't simply plug the instrument into the spy plane as we might screw a light bulb into a socket. The U-2 was a delicate bird, a truly minimalist airplane. Back in the 1950's, when Kelly Johnson of Lockheed designed the U-2, he wanted to make it fly as high and as long as possible on a minimum of fuel (a special kind that wouldn't boil at high altitude), so he cut the aircraft to the bone, removing every unnecessary strut, girder, bolt, you name it.* The U-2 was held together mainly by its thin metal skin—stiffened (like a balloon) by air pressure. To provide a viewing port for the DMR, we couldn't just cut a hole in the U-2 roof; that would weaken the metal skin, causing the aircraft to become unstable in flight. It might even break apart; the pilots told us plenty of horror stories about U-2s that were flying fine one second, then shredded to bits in the next. We joked there was a simple way to avoid all these problems and still get our astronomical data: fly the U-2 upside down. Fortunately, Lockheed's Bill Ferguson solved our problem when he discovered that the Air Force had already secretly designed an upper viewing hatch for the U-2 in order to monitor the reentry of test intercontinental ballistic missiles, or ICBMs. Few people knew about it. After calling in a lot of favors, Bill Ferguson was able to acquire the upper hatch for our experiment.

With that headache behind us, Hal Dougherty, our master mechanic, made a full-scale mock-up of the U-2 payload bay. That way we could fit all the parts together in our lab. Hal's ability to visualize a gadget, then to make a model of it, was uncanny—like Einstein's ability to visualize what it would be like to travel on a beam of light. Hal solved problems in a day that other engineers would have spent weeks or months on, costing time and money. Our funds weren't abundant, so we were glad

*It didn't have fuel tanks. The fuel was just stored in the wings; it was the first wet-wing aircraft. The fuel made the wings heavy; on the ground they drooped. Each wing was supported by special wheels, called pogos, that fell off during takeoff.

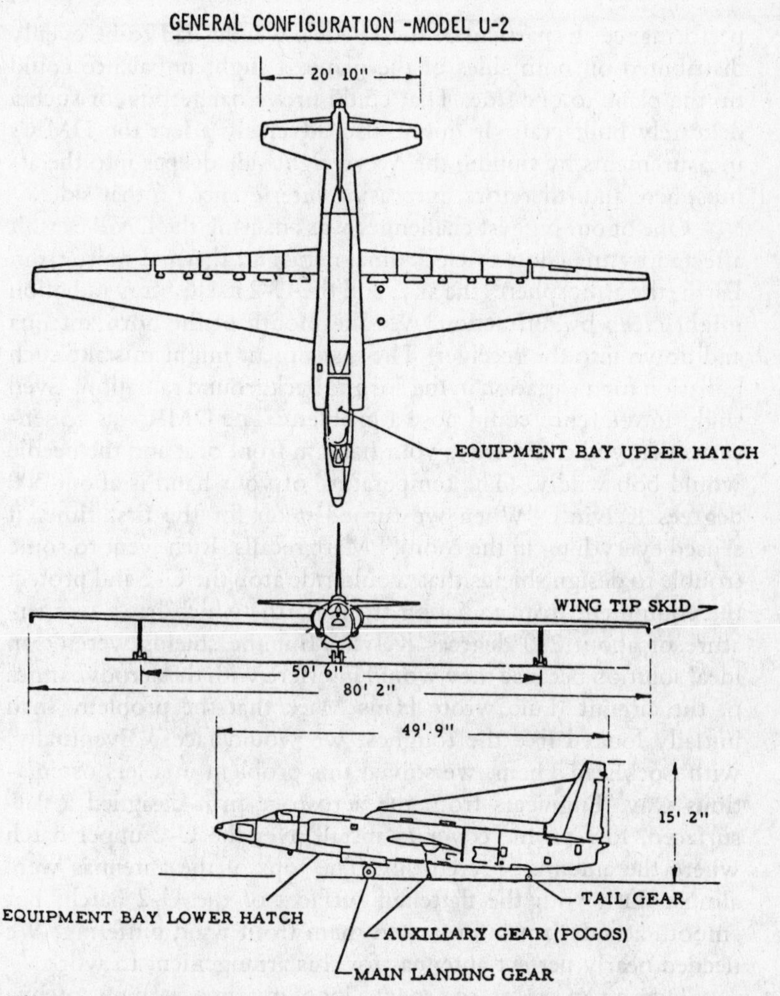

The general configuration of the U-2 airplane from *U-2 Investigator's Handbook* (1976).

that Hal's imagination allowed us to stay financially solvent.

The DMR would have to be totally automated, as the pilot would be busy flying the aircraft, and there was no room for a passenger. We also had to ensure the DMR didn't endanger the U-2 and its pilot. The DMR couldn't emit signals that interfered with the pilot's radio; also, it couldn't be too heavy, lest it threaten flight performance. In particular, the DMR's weight had to be evenly distributed on both sides of the plane; a slight imbalance could tip the plane to one side. That could prove dangerous for such a delicately built craft. It might also adversely affect the DMR's measurements by tipping the overweight side deeper into the atmosphere and, therefore, increasing interference on that side.

One of our biggest challenges was ensuring the DMR wasn't affected by the choir of electromagnetic and thermal noise from Earth, the atmosphere, the sky, and the U-2 itself. Stray radiation might creep by diffraction over the mouth of the horn antenna and down into the receiver. The instrument might mistake such radiation for a variation in the cosmic background radiation. Even slight interference could pose a problem: The DMR was so sensitive that you could wave your hand in front of it and the needle would bob wildly. (The temperature of your hand is about 300 degrees Kelvin.) "When we turned it on for the first time, it sensed everything in the room!" Marc recalls. Rich went to some trouble to design shields that would ride atop the U-2 and protect the equipment from radiation from Earth, which has a temperature of about 290 degrees Kelvin. But the shields weren't an ideal solution because they would interfere with the aerodynamics of the aircraft. Luie wrote Hans Mark that the problem "had initially looked like the toughest we would face." Eventually, with Lockheed's help, we solved this problem in a less ostentatious way. Engineers from the aerospace firm designed a flat-surfaced, low-profile cover to install over the U-2 upper hatch where the antennas peered out. The tops of the antennas were almost flush with the flattened surfaces of the U-2 hatch, but smooth aerodynamics protected them from wind buffeting. We needed nearly perfect antennas for this arrangement to work.

I wasn't an expert, so I spent a lot of my time learning antenna technology and theory. I traveled to TRG Corporation in Boston (now owned by Alpha Corporation) to meet the company's antenna engineers and to work out the specifications and design for the antennas. We chose a beautiful pair of thirty-three gigahertz (GHz)* antennas, because at that frequency there's a minimum

*Thirty-three GHz is a frequency of 33 billion cycles per second and corresponds to a wavelength of light equal to 0.91 centimeters, or about one third of an inch.

of atmospheric microwave emissions, thus giving us a less impeded view of the cosmic background radiation. The horns' directionality (their ability to see straight with a minimum of input from the sides, like a human being with extreme tunnel vision) was optimized using a technique called apodizing. This involves cutting grooves (corrugations) on the inner wall of the antenna; they are parallel to the mouth of the horn. The grooves block most off-axis radiation as it tries to sneak inside the horn and travel down its wall.

We continued to worry about the U-2's stability and symmetry. Although the aircraft was stable enough to take beautiful reconnaissance pictures, we didn't know if it would be stable enough for cosmological research. It would have to fly exactly level with the horizon; the slightest tilt would cause one horn to view through more or less of Earth's atmosphere and, hence, through a region of more or less microwave interference. To correct for this, we decided to have the pilots follow flight paths in the shape of long, thin racetracks. The plane would take off and fly in a certain direction on a long straight path, then quickly bank into a 180-degree turn and fly back in the exact opposite direction, along a path precisely parallel to the outgoing track. That way, any false signals caused by plane tilting or asymmetry would tend to cancel out.

Despite this precaution we remained paranoid about possible atmospheric interference, so we added another safety measure: an extra radiometer, the 54-GHz DMR. The atmospheric microwave emission is about four hundred times greater at 54 GHz than at 33 GHz, and so the new radiometer would detect the slightest fluctuation in atmospheric signals and, hence, the slightest tilting by the U-2. After the flight, we could compare the 54-GHz signal with the 33-GHz signal to distinguish between signals caused by cosmic anisotropy and those caused by the atmosphere. (The 54-GHz DMR worked wonderfully, as it turned out: It measured the plane's tilt with a precision of better than one sixth of a degree.)

We also had to work overtime trying to eliminate stray heat sources from the horns. But we couldn't eliminate them all, so we resorted to a secondary solution, one similar to the racetrack paths technique. It was an "if you can't beat 'em, join 'em"

Around the radiometer used aboard the U-2 flights are, *left to right,* Luis Tenorio, Charles Lineweaver, John Gibson, Giovanni De Amici, George Smoot, and Jon Aymon. All were members of the COBE DMR group from Lawrence Berkeley Lab. LAWRENCE BERKELEY LABORATORY

solution, except you could call it "if you can't beat all heat sources, make sure both horns share equal exposure to the heat." That meant the horns had to be as similar as possible, technologically speaking, so that neither would be more or less "deluded" by false signals than its partner. (Equal delusion was okay; unequal delusion was bad.)

Development of the radiometers was a lot of work, and everyone burned the midnight oil for weeks or months at a time. Personal lives suffered. The slow pace of progress sometimes depressed us all. Marc Gorenstein recalls one day driving down the hill from the Lawrence Berkeley Laboratory in his beat-up '68 Chevy Malibu. It had been a bad day; the DMR simply refused to work correctly. He conceded he was "very discouraged." And then he came to the bottom of the hill and pulled onto Oxford Street, where students relax on a strip of green grass, and noticed cars were parked all along the side of the road. He thought, Look at those cars. I could go to almost any of them, turn the key, and it would *work*. It would start up and carry me down the street. Yet each one is a very complicated piece of machinery.

It was a simple but important moment, a plain statement that every day people get thousands and thousands of very complicated machines to work. It seemed like an epiphany, and it renewed Marc's faith in the DMR.

Our morale was most boosted by the U-2 pilots themselves. These were proud, capable men, and we couldn't have put our DMR in more trustworthy hands. "Who's the best U-2 pilot I ever knew? . . . Me," declared Jim Barnes. Then he laughed. "Of course, everyone else here thinks they're the best."

We got to know the pilots pretty well. For example, Ivor "Chunky" Webster was a droll Britisher who fit the image of an old RAF pilot; you could imagine him dodging Nazi bullets in the skies over wartime England. He came to the U-2 program as part of an exchange program with the Royal Air Force. Chunky was a hearty type, but also the kind of pilot who seemed to have a sixth sense about his plane. One day, years earlier, he had been flying a U-2 at Edwards Air Force Base. Kelly Johnson, the U-2 inventor, was on the ground monitoring the flight. Chunky radioed the ground that the plane wasn't handling quite right; he sensed one of the wings was slightly bent. Experts on the ground were skeptical. Someone asked Johnson about it; he declared, "If Chunky says it has a bent wing, it's got a bent wing." After Chunky landed, technicians checked the plane. The wing was ever so slightly, barely perceptibly, bent. That's how good Chunky was.

The pilots had a complex code of honor; one rule was that they were never, ever supposed to let their wingtip skids touch the ground on landing. "Boy, if you let a wingtip skid drag, you got ragged unmercifully," Barnes said. "Or if you did something unforgivable like blow out one of those little bitty three-hundred-psi tires, whew! Bad deal! You had to buy a case of beer for the crew that had to change that tire." In a third of a century of flying the U-2, he blew a tire just once.

Barnes recalled how, in the 1950's, while he was a USAF pilot, he was asked if he was interested in joining a secret project.

Initially, he was given no details. The brass courted him at meeting after meeting, feeding him information in fragments, cautioning him that this was a top-secret project, ultrasecret. They made clear the dangers: "You're going into a hostile environment that medical science knows nothing about—except we know it's hostile. You'll fly a new kind of airplane that is very fragile. The life-support systems aboard the plane are all new. It's never been done before. You're going into the world of the unknown." Was he absolutely sure he wanted to take on this responsibility? They advised him: "Go home and sleep on it." He didn't have to; he was in all the way, and said so. Based on what he had already learned, "it sounded so exciting I couldn't stand it. I hate to say this, but I didn't consider the risk. I never did."

The same thing happened to Bob Ericson. In 1956 he was an air force pilot at Tinker Air Force Base in Oklahoma City. He was unmarried and was making five hundred dollars a month. Joe McCarthy would die soon; Stalin had died three years earlier. But the Cold War was going strong; soon the Soviets would send tanks into Hungary and launch *Sputnik*. The United States was desperate to know what was happening behind the Iron Curtain, especially within the USSR's supersecret weapons laboratories. One day Ericson was called into his commander's office. An official told him the Pentagon was developing a "special program" outside the regular air force; would he be interested in joining? "Yeah, I would," Ericson replied. Over the next few weeks and months, he was invited to top-secret briefings. At each briefing he was given a little more information about the "special program." It had something to do with reconnaissance . . . monitoring Soviet tanks and troops . . . a new kind of airplane . . . that kind of thing. It was like watching a jigsaw puzzle being assembled, piece by piece. Occasionally, the briefing officer urged Ericson to go home and think about what he was doing. Was he sure he still wanted to participate? He assured them he was, despite the risks.

Like all pilots, Barnes and his colleagues loved to tell stories, the kind that hold an audience in disbelief, horror, and admiration. Listening to such stories, we were amazed anyone ever came back from a U-2 flight. As they described it, engines would "flame out," or the plane would fly a little too fast and suddenly fall apart.

Barnes recalled a day years earlier at the U-2 training base in the Mojave Desert. One of his buddies, a man named Buster, was wheeling overhead in the U-2. Buster radioed ground officials that something was wrong with one of his wings. "I'm going to take this thing into slow flight to see what's wrong with this wing," he told them. Barnes says Buster then soared "to about two thousand feet. We're staring at him, and I said to the guy next to me, 'Huber, he's pulling the power off.' We're watching when all of a sudden, while Buster is at two thousand feet, his left wing turned down and snapped off. Instantly! He went into a big spin, made about one and three quarter turns, and down he went, making a sound like *zoooooommmmm—smack!* Right before my eyes. Unbelievable. He wasn't the first one I watched die."

Ericson recalled late 1962, when the world teetered on the brink of nuclear war. His U-2 photos were the first to reveal the presence of secret Soviet missiles on Cuba. That was before President Kennedy had publicized the missile shipments. "The only thing my photos did for me was to get me grounded for three or four months. My bosses thought, Gee, Ericson knows too much—we can't let *him* get shot down." His photos also showed thin white lines arcing upward. Photo interpreters told him they were Cuban missiles rising from their launch pads toward Ericson's U-2. The Cubans had been shooting at him, and he hadn't even known it.

Barnes dismissed as an "old wives' tale" the legend that U-2 pilots carried a cyanide pill to swallow in case they were shot down. But they carried something else. He slowly, dramatically pulled a coin from his pocket, a silver dollar in the side of which was bored a hole the diameter of a pencil lead. The hole went through the coin and ended about a sixteenth of an inch from the far side. "You unscrewed this and you pulled out a small sheath, which covered a very fine needle that was covered with the most deadly substance known to man. A superconcentrated curare." Native tribes used curare for deadly darts; the silver dollar contained curare that was two thousand times as potent. With it,

Barnes said, a downed U-2 pilot could kill a sheep or other animals for food. Or he could kill an enemy soldier. Or, if there was no escape, "you could use it for something else."

Finally we were ready for a full-scale test of our instrument, the newly minted differential microwave radiometer. On a night two weeks before the first flight, Marc Gorenstein and I loaded the DMR on a wooden cart and put it in the back of our pickup truck, *Old Yellow*. We covered the instrument with a tarpaulin and secured it with chains. Then we drove through the darkness to an isolated parking lot at Lawrence Berkeley Laboratory. There, we rolled out the DMR and pointed its twin horns at the dark sky. Our meter bank showed us what the DMR was seeing. An invisible patch of water vapor drifted overhead; the scanner showed a rise in temperature. Good; this meant the instrument was working, because water vapor was a source of stray radiation. The passage of such clouds was a reminder of why it was so important to get the DMR into the stratosphere, far above clouds.

We also unloaded a special pair of big polished mirrors, made for us by Hal Dougherty. We placed them at angles so they would reflect radiation from directly overhead into each horn antenna (which were pointing sixty degrees apart). If both antennas were looking at the same patch of sky through identical mirrors, then the horns should detect a temperature difference of zero—if they were working properly. As it turned out, the horns saw a temperature difference close to zero, which wasn't close enough for our needs. So we took the DMR back to the lab and adjusted the two sides of the radiometer to make them more symmetric. We took the DMR out again the next night for another test. After two tries it passed perfectly; both horn antennas saw the same amount of radiation.

A few days later we put the DMR in the truck and drove the fifty miles from Berkeley to Ames. There, we tested our creation aboard the U-2, while it was still in the hangar. Marc and I took turns as the "pilot," sitting in the cramped cockpit and operating the DMR by pressing various buttons. Meanwhile, the other one would place microwave-emitting targets (sources of different levels of radiation power) in front of the horns to test their sensitivity. When everything checked out, we let the real pilot taxi the U-2 down the runway with the instrument aboard. As part of the test, he switched on his radio and other equipment, so we could see if they would interfere with the DMR. No problems were detected. We were ready for flight.

Our first U-2 launch was set for July 7, 1976, three days after the Bicentennial and two weeks before the first unmanned Viking probe was to land on Mars.

It was early evening; the sun disappeared behind the coastal mountains to the west, and the blue sky turned dark. The U-2 pilot Ron Williams got into his space suit, donned his helmet, and walked out to the plane, looking like an astronaut headed for the Moon. In fact, his mission (via the DMR) was far beyond the Moon, far beyond our galaxy, and deep into the cosmos. He squeezed into the cockpit, closed the transparent hatch, and went through the standard checkout. He may have said a silent prayer; I don't know. He revved his engine and crept down the runway, supported by detachable wheels, his wings drooping with the fuel load.

Then he accelerated and took flight. The extra wheels fell off. Suddenly the plane seemed to angle almost straight up, like a rocket. We gasped at the sight: One second the U-2 was level with the ground and the next we were watching its tailpipe, white flame against the indigo sky. By the time he passed over the Dumbarton Bridge, he had thirty thousand feet under him.

In the stratosphere, the pilot switched on the DMR. It began rotating back and forth. He could tell it was working because his instrument panel included a little light that blinked on and off every time it rotated. The horns switched position every thirty-two seconds, automatically and repeatedly comparing the temperatures of distant parts of the cosmos.

Soon, two years of intense team effort would be put to the test. No instrument like this had flown in the stratosphere before. We hoped our hard work would pay off and nothing would go wrong. On the ground, I analyzed the initial results as they came in. I wasn't sure what we would find, but whatever the data showed, I felt sure it would be interesting.

Chapter 7

A Different Universe

Within a couple of months of that first U-2 flight in July 1976, it was clear our experiment would work. The data from the differential microwave radiometer began to flow. We had stringently tested the system in several engineering flights and had moved into what we called scientific flights. The U-2 plane proved to be phenomenally stable, flying level to about one sixth of a degree on the average, just what we needed to minimize false signals. Several teething problems with the DMR were quickly fixed. Then it embarked on its "scientific" flights, with their demanding task: measuring minute differences in the cosmic background radiation in different parts of the sky. While these flights were under way, we managed to boost the DMR's sensitivity and developed computer programs for analyzing the data. We were poised to use the cosmic background radiation—that dim afterglow of the immense energy of creation—to explore the dynamics of the universe.

At this time—the late 1970's—"classical" big bang theory viewed the origin and evolution of the universe as a relatively orderly affair. Following the primordial event 15 billion years ago, gravity caused galaxies to slowly condense on small regions of higher-than-average density (cosmic seeds) in an otherwise uniform medium. Space, it was thought, was sparsely populated in the extreme with these gossamer galaxies, which proceeded through time relatively unperturbed by other heavenly bodies. Any part of space was much like any other—it was all relatively homogeneous. We knew that the gravitational seeds for these structures must have been planted early in the history of the universe, slowly to develop thereafter. The assumed relative uniformity of galaxy formation and distribution, however, did not inspire urgent inquiry into the nature of these seeds (this was before the discovery of gigantic superclusters of galaxies). So I was more interested in addressing some of the Big Questions about the dynamics of the universe: Did it rotate? Did it expand uniformly in all directions?

We didn't have to wait long for answers. With half a dozen scientific flights completed, it began to be all too clear to us that the DMR data contained no hint of rotation of the universe. This was a major surprise, because we can see that everything *within* the universe is rotating—planets, stars, and galaxies. I had convinced myself that the universe should be rotating. All systems, from the proton up to galaxies, rotate, so it made sense that the universe would rotate, too. I knew that general relativity allowed rotation in spite of the inevitable question: What does the universe rotate with respect to? From our results, we calculated that if the universe does rotate, it had done so at less than one hundred-millionth of a rotation in the last billion years. At that level we had improved the accuracy of the Mach's principle tests by a large factor. We could improve still more, by finding the new aether drift caused by our galaxy's rotation But lack of rotation by the universe meant time travel was going to be difficult.

We also found the universe is expanding with remarkably uniform speed in all directions. There was no hint of asymmetry. "The big bang, the most cataclysmic event we can imagine, on closer inspection appears finely orchestrated," I wrote at the time of our observations. "Either conditions before the beginning were very regular, or processes we don't yet know about worked to make the universe extremely uniform." This conclusion was innocuous enough, and would upset no one, as the uniformity of the universe could be seen as being consistent with classical big bang theory. It was only later—together with other discoveries—that it would be perceived as a problem.

What did upset people, however—particularly astronomers—was a completely unexpected discovery. We stumbled onto evidence that the universe was not at all as had been imagined.

The Milky Way, like all spiral galaxies, rotates, a giant disk of stars slowly turning as its stars orbit around its center. We knew that we should see a signal of this rotation in the cosmic background radiation. Because of the Doppler effect caused by our Sun's motion around the Galaxy, the background radiation would look a little warmer than average in one part of the sky (the Sun's direction of progression) and a little cooler in the opposite part of the sky (the direction of regression). We also knew that this temperature difference—known as dipole anisotropy—would be small. If our data failed to reveal the dipole, we would know there was something wrong with our instrument or our analysis.

In early December, five months after our first flight, I felt we were at the brink of finding the expected dipole signal. Jon Aymon and I worked long hours, fighting with a balky computer-operated tape reader on more than twenty-one hours of U-2 observations on data cassettes. We had to reject only six minutes of data because of contamination by various forms of interference. That illustrated how well the system worked. Now we had all the data in the Lawrence Berkeley Laboratory's big number-crunching computer, which spit out summaries of each set of racetrack runs. After we processed and checked each flight's data, I added it to the previous batch and ran it through my programs. There was a hint of a dipole forming in the celestial map we were cumulatively building. If that's the case, I said to Jon, we should be able to predict how the data should look from the newest flight. I made the calculations, sketched out how the points should fall on the map, and then looked at the real data. The match was extremely close.

This was exciting. But I wanted to check its significance, so I suggested Jon write a program to test the results by simulating the data with no signal—only instrument noise. If this also produced a good match, we would know our results weren't significant after all. It didn't. It looked just like what it was—noise. We therefore became more confident. We had been plotting our data on a large celestial sphere—a globe of the heavens, with the constellations, brightest stars, the outline of the Milky Way, and the celestial coordinate system marked on it. I put colored marks on the globe indicating the temperature variations. We worked through the night on this, and by dawn we could see the pattern we were looking for: the dipole anisotropy, showing part of the sky warmer (blueshifted) and the opposite part cooler (redshifted).

The degree of anisotropy was small—just one tenth of 1 percent (one part in a thousand). However, all was not as expected. "Look at that," I said to Jon. "What do you suppose that means?" Although the anisotropy was close to the magnitude we had expected, its direction was nearly the opposite. That is, the sky was warmest in the direction of Leo and coolest in the direction of Aquarius, which means that Earth was moving toward the former and away from the latter. That is *not* the direction in which the Galaxy rotates. "Unless we have a sign wrong," said Jon, "there's only one explanation." We both knew what the answer had to be: Not only is the entire Galaxy rotating, as it should be, but, unexpectedly, it is also moving through space. And it was moving very fast—six hundred kilometers a second, or more than a million miles an hour.

Jon and I were exhausted by morning, but excited by what we had found. We could barely wait to share the news with the rest of the team. The implications, we knew, were profound.

Think of it this way. When Copernicus announced Earth was orbiting the Sun, someone surely sat down with a pen and paper and said, "Okay, let's do some calculations. If Copernicus is right, and Earth has a nearly circular orbit around the Sun,* then, according to the formula for the circumference of a circle, the orbit must be . . ." scribble, scribble ". . . almost 600 million miles in circumference. And since a year has 365 days, or about 8,800 hours, or about 530,000 minutes, or about 32 million seconds, therefore . . ." scribble, scribble ". . . Earth is orbiting the Sun at a speed of approximately 19 miles (30 kilometers) per second. Nineteen miles per second! Why doesn't everything fall off it? The guy must be nuts!"

That is a sixteenth-century version of what we found, which came to be called the "peculiar velocity" problem. Copernicus's

*In fact, Earth's orbit is slightly elliptical.

contemporaries were amazed to learn that Earth moves at all. Likewise, in the 1970's, many of our colleagues would be incredulous that our galaxy should be moving through space at such a high speed relative to the cosmic background radiation. Theorists would be prepared to accept a galactic motion of, say, sixty kilometers a second—but not something greater by an order of magnitude. The only way our results could be correct was if there were a gigantic celestial mass out there—unseen and unsuspected—that was dragging the Milky Way toward it under the influence of gravity. According to prevailing theory, matter is distributed fairly homogeneously through the cosmos, and no such bodies of extraordinary size were posited to exist anywhere. In a paper destined for the *Physical Review Letters,* we wrote: "The large peculiar velocity of the Milky Way galaxy is unexpected, and presents a challenge to cosmological theory. . . . The Universe may be much more inhomogeneous than we have realized until now." If the result was correct, we were certain we would stir things up.

One thing worried me, however: Could the dipole have been a seasonal effect? When Earth is on one side of the Sun, say in spring, it moves one way relative to the Galaxy, and when it is on the other side (autumn) it moves in a 180-degree opposite direction. So how could we be sure the dipole wasn't simply an artifact of the seasonal change, and unrelated to galactic motion through the cosmos?* We checked the potential seasonal problem by asking the U-2 crews to fly additional flights, on which the DMR could rescan parts of the sky observed during flights earlier in the year. This required the pilots to work flat out, flying much later at night than usual. When we merged the new DMR data with the original set to correct for any seasonal effect, we held our collective breath. To our relief, there were no seasonal changes; the dipole was still there.

Confident that our data were valid, I planned to announce our results at the American Physical Society meeting in Washington, D.C., in April 1977. Our competitors, particularly the

*The 1991 report by English scientists that they had detected a planet orbiting a distant pulsar proved to be wrong because they had failed to account for Earth's motion around the Sun. They had detected planet Earth.

Princeton team of Brian Corey and David Wilkinson, were hot on the track of the dipole with balloon-borne observations, but I wasn't sure how close they were. It was important to us to get our work on the record, and the APS gathering provided that opportunity. We weren't able to get a regular place on the speakers' agenda because we weren't convinced of our results until after the deadline for submission of abstracts and invitations. Luckily, Jim Peebles, whose book had inspired our U-2 experiment, was giving an invited talk on cosmology and generously yielded a few minutes for me to talk. It was somehow poetic that he would be there, as if to complete the cycle of the idea of the new aether drift experiment.

Before the meeting, I had practiced the short talk about ten times. This was a big day for me; I had dressed in a coat and tie, which was not my usual garb. My girlfriend, Constantina Economou, came to cheer me on.

Standing before my colleagues in the large meeting room, I felt young and nervous. Peebles made a kind introduction and

George Smoot (*left*) and Jim Peebles together at a conference in 1992. EL ESCORIAL—CONVERSACIÓNES DE MADRID

offered me the podium. I went through my talk quickly and explained that, while Earth and the Solar System are moving toward Leo at about 350 kilometers per second—more than ten times the velocity of the Earth going around the Sun—the Milky Way galaxy is traveling about 600 kilometers per second. "This is quite a large volume to be moving with such large velocity," I said wryly. I remember thinking: I bet Jim is the only one here that really understands what this implies. Then my talk was over; I fielded three questions, all straightforward, and before I knew it Peebles was talking again. Our detection of high-speed galactic motion presented "a real dilemma for the theorist," he said almost nonchalantly.

That was an understatement. The audience that day was composed mostly of physicists, to whom such extraordinary movement of physical objects seemed natural and acceptable. Our results were only on the fringe of their interests, and therefore challenged no cherished worldviews. The reaction from astronomers would be different.

Six months after the physicists' conference in Washington, Marc Gorenstein presented our results to an American Astronomical Society meeting in Atlanta. The audience's reaction was cool, not to say frosty. Marc was closely questioned, with the clear implication that there must be something wrong with either our measurements or their interpretation. The implications of the results, as we saw them, so threatened prevailing astronomical wisdom that few astronomers were willing to take us seriously.

Marc's presentation caught the eye of Walter Sullivan of *The New York Times,* who wrote a story about our work on the November 14, 1977, front page, under the headline GALAXY'S SPEED THROUGH UNIVERSE FOUND TO EXCEED A MILLION M.P.H. In it he succinctly explained the challenges our results posed to orthodox cosmology. First, the extraordinary motion of our galaxy demands the existence of massive, previously undetected bodies in the universe, which means that the universe is not nearly as homogeneous in its distribution of matter as had been thought. Second, this in turn raises to a new level of interest and importance the means by which the present universe evolved after the big bang. The unexpected existence of enormous structures in today's universe meant that the cosmic seeds from which they grew must have been present in the very early universe. Otherwise the structures could not have grown so large by now. One effect of our results, therefore, was to focus attention on the nature of these seeds and how to detect them.

The astronomers attacked us on two fronts: empirical and theoretical. Canadian astronomer Robert Roeder of the University of Toronto pointed out that our cosmic background radiation map was aligned closely with the plane of the Solar System. How, he asked, could we be sure the "anisotropy" wasn't just dust orbiting the Sun? Dust such as that producing the zodiacal light, which is caused by particles left over from the collisions of asteroids. Roeder also noted that, for the period December through May, the signal seemed to be in the same direction in which the zodiacal light was brightest.

He had a point. Before releasing our U-2 data, I thought I had adequately accounted for any effects caused by the zodiacal dust. But you never know; perhaps the Solar System contained bigger dust particles than we had considered, particles that might have caused the temperature differences we detected.

In subsequent months Roeder sent me letters in which he expanded on his original criticism. He analyzed our data and confirmed that there were unsettling correlations between our anisotropic signal and the plane of the Solar System. (Almost all matter in the Solar System, from dust to planets, orbits in roughly the same plane.) Our signal lay within five to ten degrees of the plane of the zodiacal light. Also, the warmest part of the anisotropic signal was right on the plane of the ecliptic (the path the Sun takes across the celestial sphere). Furthermore, the anisotropic signal was strongly aligned with the point where the ecliptic crosses the celestial equator (which lies on the same plane as Earth's equator). Maybe these were just coincidences. Even if they were, those are the kinds of coincidences that give astronomers sleepless nights. I had no definitive answer, and it was only later that we were able to eliminate Roeder's objections.

I must have spoken about our findings at about a dozen meetings; Marc and Rich also gave many talks. Practically every time, critics expressed doubts about our measurement of galactic velocity. Even some local Berkeley astronomers doubted the existence of the dipole. The theoretical objections stemmed from the fact that our findings challenged a central axiom of modern cosmology: the Hubble law. According to this law, established by Edwin Hubble in the 1920's, the steady expansion of space is responsible for the linear correlation between the distance to a galaxy and the speed of its recession. A galaxy might have a slight peculiar velocity independent of cosmic expansion, but this velocity was assumed to be very slight indeed, as various experts had assured me prior to our U-2 experiment. How could galaxies be moving as fast as we claimed and still abide by the Hubble law? That is, if galaxies tend to move at extremely high speeds in random directions, then there should be much more "scatter" from the central line on the Hubble diagram—the relationship could not be as tight as had been held.

The implication that our galaxy was being pulled by an inconceivably huge and distant mass was, of course, a major problem, because no theoretical model of the universe included such structures, and none had been seen. If such a massive gravitational source existed, then why didn't it tear our galaxy apart? Astronomers had cataloged numerous galaxies that had been disrupted by close encounters with other galaxies. Such cataclysms demonstrate that a galaxy is like a bag of feathers without the bag: Perturb it, and it falls apart. In the 1970's, it was hard to grasp why a gravitational pull powerful enough to accelerate the Milky Way to six hundred kilometers per second would not also disrupt it by tidal effects. And the Milky Way was not alone in its extreme velocity. About a dozen neighboring galaxies—the Local Group—are also moving presumably under the influence of the distant, unseen structure. Producing this speed without disrupting the Local Group would require a very large mass concentration very far away; that would permit the change in gravitational pull over millions of light-years to be small (small enough to avoid disrupting the Galaxy). Years later, other astronomers launched a serious search for such a structure and found it—"The Great Attractor." Large as it is, it is unlikely to be unique. There are probably many more in the universe.

In any case, in the 1970's there was no theory to explain how such massive gravitational sources could have formed. The primordial seeds, or wrinkles, that gave rise to structures in the universe were at that time thought to have been comparatively small—producing weak gravitational attraction—hence, the resulting galactic clusters must be comparatively small, too.

These were all potent objections and we had no way to answer them. To some, we were coming dangerously close to challenging the foundations of modern cosmology. Scientists are understandably reluctant to accept extraordinary ideas in the absence of extraordinary evidence, and at the time many astronomers didn't consider our evidence to be extraordinary. Looking back on it, I don't fault them for their caution. They were doing to us what Luie Alvarez always did to us, his students, when we claimed to have detected this or that effect: He grilled us, often brutally, to ensure we had dotted every *i* and crossed every *t*. In science, such grilling is essential. How far would the "cold fusion" debacle have gone if physicists hadn't thoroughly scrutinized the initial claims?

In any case, we weren't the first astrophysicists to be grilled for challenging the orthodox concept of cosmic homogeneity. Vera Rubin had beaten us to that honor—by a quarter of a century.

On a snowy day in December 1950, the twenty-two-year-old Rubin and her husband were driven by her father to a meeting of the American Astronomical Society at Haverford College in Pennsylvania. She had given birth to her first child three weeks earlier. A Vassar graduate then working on her master's thesis under George Gamow at Cornell University, she had never been to an AAS meeting before. Now she was scheduled to speak at one. Being "a dumpy graduate student who didn't have anything to wear," as she mockingly describes herself, she had bought a blue knit dress for the occasion. When she entered the conference room, she anxiously scanned the audience. She didn't know a soul. Her father and husband seated themselves. She began to speak. Her talk had originally been titled "Rotation of the Universe," but the meeting organizer thought that sounded odd, and so he had changed it to "Rotation of the Metagalaxy."

Rubin posed a seemingly simple question: Does the Hubble flow adequately explain how the cosmic density of galaxies changes over time? As space expands and galaxies "move" apart—just as bugs sitting on a balloon "move" apart as the balloon expands—do any galaxies display additional motions unrelated to the expansion? In retrospect, it was a reasonable question. Even Earth's rotation isn't a simple rotation about a north-south axis; the planet also undergoes subtler, cyclical "wobbles" that repeat over thousands of years. (For example, because of a wobble known as precession of the equinoxes, the so-called North Star, or Polaris, is cyclically displaced from, then returned to, its perch as the point around which the northern sky rotates.) Likewise, Rubin wanted to know if galaxies have motions independent of the Hubble flow. It was like asking if bugs on a balloon move

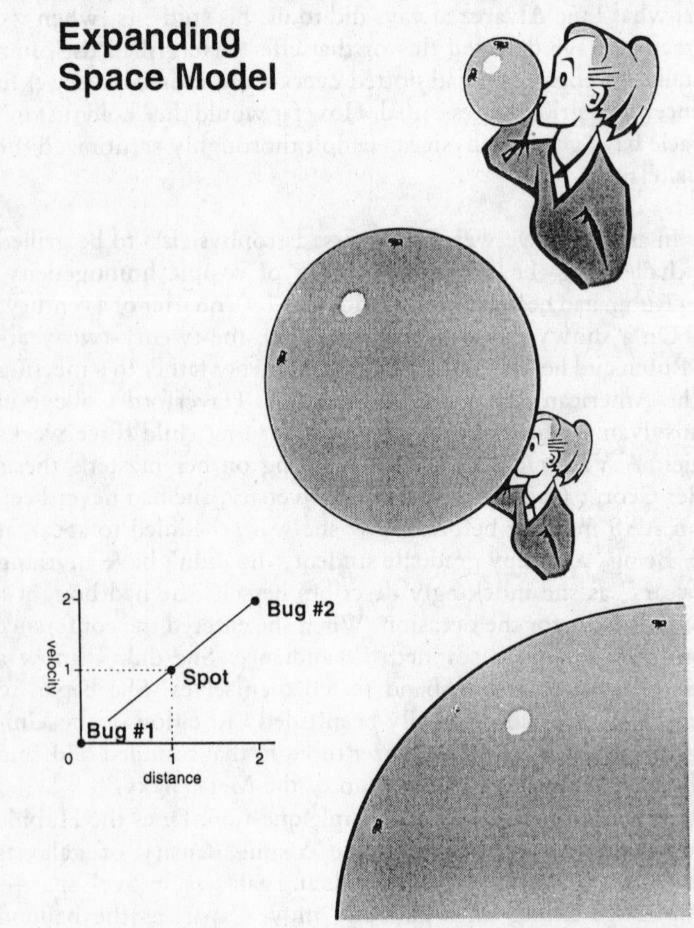

on their own across the balloon surface as it expands.

To find out, she analyzed more than a hundred galaxies to see if they were receding at the exact speed predicted by the Hubble flow. According to Hubble's law linking redshift with distance, a galaxy's redshift—which reveals its rate of recession—is proportional to its distance. So Rubin compared the galaxies' redshifts (measured by other astronomers) with their distances (which she estimated according to their brightness—a risky maneuver, based on the unproven assumption that all galaxies of the same type have similar absolute brightness). Her goal was to see if galaxies moved independent of the effect of the expansion of the universe. She found they did. Some galaxies were moving much faster or slower than they should if the Hubble law were ironclad, she told the audience. (Likewise, we later found our Milky Way galaxy is moving far faster than previously accepted, at six hundred kilometers per second in a direction angular to the Hubble flow.) This was surprising news. Rubin's talk was daring; it implied that the Hubble law wasn't sacrosanct and that the Hubble flow was far from smooth.

Her talk was "very, very badly received. . . . There was a real hubbub," she recalls. People were particularly upset because her analysis didn't include error bars. Error bars represent an estimate of the error in the data, and thus are an indication of the range over which the results would be expected to vary with repeated measurement. Modestly, she recalls she didn't know at the time how to prepare error bars for such a complex analysis. "They felt my analysis was shaky—which it undoubtedly was," she admits. To calm the waters, the distinguished researcher Martin Schwarzschild stood and said the anodyne things that senior researchers sometimes say to young students in such circumstances. He said: "This is a very nice paper. There may not be enough data, but it was an interesting thing to do." A critic persisted by asking her, "Why didn't you get error bars?" "Well," she replied, "the entire story is so . . . nebulous." There was a roar of laughter over her inadvertent pun. She laughed, too. "So it ended on a pleasant note."

After the coffee break, the Rubins drove home. "I didn't feel bad at all," she recalls. "I had given a very good talk; I had

Vera Rubin, whose graduate work indicated that galaxies have peculiar velocities. CARNEGIE INSTITUTE OF WASHINGTON

virtually memorized it; I [still] could have given it ten years later. I went home and nursed my son."* Newspapers picked up the story. *The Washington Post* had a front-page article with the headline YOUNG MOTHER FINDS CENTER OF CREATION.

Astronomical journals declined to publish Rubin's paper, and her work was almost totally ignored by astronomers. The Texas-

*Decades later, Rubin encountered an astronomer who had served on the AAS committee that chose speakers for the 1950 meeting. The committee had had a "very serious discussion" about whether to let her speak, he told her.

based French astronomer Gérard de Vaucouleurs, however, took her seriously. He was interested in the possibility that the universe might be less homogeneous than generally thought—that is, that galactic density might be much greater in some places. Such dense regions might exert enough gravitational pull to attract some galaxies at the speeds reported by Rubin, he suspected.

The idea of clusters of galaxies wasn't new. For example, in the 1930's, Fritz Zwicky, a Swiss astronomer working at CalTech, had suggested that galaxies might gather in clusters as large as 10 million light-years,* and that "cluster cells" might span the universe in the way "suds divide a volume of suds." But de Vaucouleurs was talking about much bigger clusters than envisaged by Zwicky. As de Vaucouleurs later acknowledged, Rubin's work inspired him to publish his data indicating that there was a "stream" of galaxies across the northern galactic hemisphere. Galaxies in the so-called Virgo cluster were largely situated within a huge disk, which he dubbed a supercluster. "It occurred to me that this belt of galaxies had to be a real physical association," not a coincidence, de Vaucouleurs said during an oral interview with Alan Lightman in the 1980's. "I was given the incentive, or the courage, to publish this because Vera Rubin had just published an abstract of her master's thesis." Because of the hostile reception, Rubin didn't touch the subject again for two decades. But, de Vaucouleurs said, "I am more pugnacious, probably, and if it's there, damn it, I'm going to say it's there." Still, the astronomical community reacted to his work with "complete silence."

In de Vaucouleurs's office at the University of Texas at Austin, his early critics' doubts are recorded on his blackboard. Even Zwicky is quoted as saying (in 1959): "Superclustering is nonexistent."

And that was the orthodox view for years to come, the orthodoxy against which our initial U-2 results had clashed.

Among the many objections raised to our findings, one was unquestionably valid, and we had no choice but to respond. Could the extreme speed of the Galaxy be an illusion caused by a "local"

*The largest known supercluster today is at least tens of times wider.

effect, astronomically speaking? We had made observations only in the Northern Hemisphere. "If you see the same anisotropy in the Southern Hemisphere," our critics in effect said, "then we'll believe you."

So south we went. If we saw the same dipole effect in the Southern Hemisphere, then the critics would have to put up or shut up.

Since the nineteenth century, South America has become an increasingly important base for astronomers. It was there that Harvard astronomers photographed Cepheid variables—photographs that allowed Henrietta Leavitt to deduce the Cepheid period-luminosity relation, the key that unlocked the door to the expanding universe. More recently, in 1987, astronomers there discovered and studied the first major supernova near our galaxy in centuries—a supernova not visible from the Northern Hemisphere.

In a commentary, the British journal *Nature* agreed that we would be wise to conduct U-2 flights from the Southern Hemisphere, but cautioned that, given the uneasy history of U.S.–South American relations, "perhaps U-2's are still not all that welcome outside U.S. airspace." However, money, not politics, was the real barrier to our South American expedition. In January 1977 I wrote to Hans Mark and told him that I had asked NASA to fund a Southern Hemisphere trip, but that "NASA's response so far was to cut off all of our U-2 flights as of September 1977." One reason, I explained, was that the U-2s were being transferred from the control of one NASA branch to another; another reason was that we had already detected the dipole, so (officials wondered) why did we want to detect it again? "Our case is presently under review at NASA headquarters and in the meantime we are getting no data," I lamented.

It was a frustrating time. I thought we might get NASA interested again if the proposed expedition had a broader array of experiments. Perhaps we could attract another scientific team interested in sharing instrument space aboard a U-2 flight in the Southern Hemisphere. "Please inform us of any other user who might have need of Southern Hemisphere flights," I asked Mark. "Our equipment is compatible with a large number of instruments so that we could combine flights or alternate flights." This is how much of science works in the late twentieth century: One spends as much time hustling for funds and filling out forms as one spends in the lab. In our case, the hustling paid off; the money finally came through. We did share the U-2 with other experiments.

Then we faced another question: Where would we go in the Southern Hemisphere? Our first choices were Australia, Argentina, and Chile. But Australia was too far away. The problem was, the U-2 couldn't fly there directly; instead it would have had to be disassembled and flown down under in a transport aircraft, at great expense. That ruled out this option. Argentina and Chile were attractive because they were well below the equator and, hence, offered a spacious view of the southern sky; unfortunately, they were also threatening war with each other at the time. Too dicey. Peru was our fourth choice. It wasn't an ideal solution either: In 1975 the country had undergone a bloodless military coup and ever since had endured political, social, and financial chaos. Still, Peru was the best we could do.

"Surrounded by brown, barren hills, Lima sits desolately at the spot where the bleak Peruvian desert meets the sea," Milton Viorst wrote in the *Atlantic Monthly* in 1978. "A chill Pacific wind keeps it enshrouded in gray clouds most of the year. Lima is almost always damp, never rainy, so that hardly a flower or tree lightens the gloom." The explorer Pizarro had dubbed it "the City of Kings," an allusion to its glorious past, when the Andes glistened with still-untapped gold. By the time we arrived, most of the gold was gone, and the country was struggling to keep its fishing fleet solvent and its nascent petroleum industry alive. Lima had become a major, fast-growing Third World urban area, packed with five million souls.

U-2 pilot Bob Ericson recalls arriving with an advance team to check out the facilities at Jorge Chávez Airport, on the coast, near Lima. He and a colleague went first to the U.S. embassy, where they talked to the air attaché. They were told it would take "weeks" to set up arrangements for the U-2. As Ericson tells the story: " 'Come back in a couple of weeks,' the embassy guy said. So we went back to the hotel room and got drunk. We woke up in the morning with hangovers and asked, 'Well, what do we do now?' So I and this guy—Jack Wall—went up to the air base and started scouting around, and pretty soon I found the head of the operations office at the airfield. He said, 'Yeah, I've got a hangar for you. Yeah, we've got this, got that . . .' He took us over and introduced us to the head of the Peruvian navy. That was the guy the attaché was trying to set up a meeting with. We had tea or something. He said, 'Yeah, yeah, we'll accommodate this, we can do this, we can do that.' We went back to the hotel that night, got drunk again, and the next day went to see the air attaché and said, 'Hey, we're leaving town—we've got it all set up, we don't need you.' "

A single U-2 team was flown down; scientific equipment was shipped by military transport. The U-2 requires a lot of its own equipment, which has to be shipped with it. It has its own fuel, its own ground landing equipment, its own engine carts, its own nose carts, and so on. All were custom designed, so that all had to be shipped to Peru in advance. Fuel was preshipped down in fifty-gallon drums by barge and boat. The total team included about eighteen NASA and Lockheed personnel and was led by James Cherbonneaux, a former U-2 pilot and NASA's U-2 project manager. Preparations and logistics reminded me of an old-time safari or expedition.

At the Jorge Chávez Airport, commercial aircraft occupied one end of the field, military facilities the other. We set up our U-2 base near the military section, where the U.S. presence was little more than a two-room shack. It was the base for the U.S. military attaché and a channel point for diplomatic mail to South America.

On the roof of the shack we set up an automated polarimeter, which would scan for evidence of polarization (a preferred direction of vibration of radiation) in the cosmic background radiation. This experiment, which was the thesis project of my graduate student Phil Lubin, would provide an important check on the source of the dipole. If the radiation was polarized, then we would know the dipole effect is caused by our galaxy; if it was nonpolarized, then we would know it was a cosmic signal. Polarization studies work best in dry climes, and the dry terrain of Peru looked ideal for Phil's observations. Unfortunately, the measurements would be partly frustrated by high humidity, frequent cloudiness, and dustiness. By an infuriating coincidence, while we were there the region—a desert—experienced its first rain in eight years. It barely moistened the pavement, yet the resulting newspaper headlines were four inches high.

Working in Peru was nothing like working at NASA-Ames. For example, in Peru it took three weeks to get replacement parts. People got sick and lost. Worse, our equipment suffered on the journey across the equator. It is much more humid in the tropics, and as the plane descended from the cold upper air into Lima, the chilly equipment condensed the humidity into water. As a result, water collected inside the small, sensitive wave guides that connect the differential microwave radiometer's horns to the receiver. We had to take the receiver apart to dry it. While we were doing this at the airport, the lights went out—a power failure. Here we were, stuck in the dark, miserable with perspiration and standing on ladders with delicate parts in our hands. We found a flashlight, and by its lone glow we continued tinkering. We laid out the parts so they could dry, and then we left for some rest at our hotel.

Very late that night, Phil and I drove our rented VW bug back to the airport to see if the equipment was dry. To get there, we had to pass through a gate watched by guards from the Peruvian navy. I showed our pass to the guard, then drove toward the hangar housing the U-2 and our equipment. Suddenly, out of the dark came a shout: "Alto!" followed by a terrifying sound: *raaaaaaack!* Our headlights outlined a seventeen-year-old marine aiming a machine gun straight at us. I stamped on the brake. No one had warned him we were coming. (Troops tend to be nervous in a country where coups d'état are an occasional part of the political process.) We showed him our pass and explained we were part of the U-2 team. He eventually waved us on. We drove to the hangar. Our equipment had dried, so we reassembled it and tested it; it worked. With any luck, the U-2 flights would soon be collecting the data that would either establish our earlier results as valid, or show that we had been naive victims of a North/South Hemisphere effect in the cosmic background radiation anisotropy.

After the U-2 flights were completed, Constantina and I took a side trip that reminded us of humanity's long fascination with the cosmos. We were eager to see the spectacular geography and the archaeological sites of Peru, so we visited Cuzco, the capital of the pre-Columbian Incas, and then outlying locations such as the fortress Sacsayhuaman. Inca engineering is truly awe-inspiring, with huge stones cut out of a quarry and fitted together so perfectly that you can't slip a knife between them. A guide scoffed at speculators who claimed "ancient astronauts" were responsible for this architectural feat. An especially amazing part of the trip took us by mountain train from Cuzco to Machu Picchu. There we saw evidence of how important cosmology was to Incan mythology, as it was to the mythologies of all early great civilizations.

When the Inca ruler Pachacuti rebuilt Cuzco in the mid-fifteenth century, he divided the city into southern and northern moieties. The ancient city was further divided into quarters, symbolizing the four directions of the empire and the four corners of the cosmos. At the heart of the city was the Coricancha, or Sun Temple, where several magnificent gold images of the Sun were placed in its innermost shrine. A celebration of the summer and winter solstices was an important component of Inca ritual.

After Machu Picchu we took a train down the *altiplano* to Lake Titicaca and visited the Uros Indians, who live on floating islands of reeds. Then we boarded a plane and flew to Arequipa and then on to Lima. At the airport, I checked our polarimeter, reset it to scan a new part of the sky, and changed the magnetic tapes, which had almost run out.

A couple of weeks later we wound up the U-2 operation and the polarimeter experiment, and returned to the United States with a mountain of data. Analysis followed. It didn't take long to confirm that the dipole was indeed of cosmic origin, not a local phenomenon; and that it was exactly as we had described it from the Northern Hemisphere. Conclusion: We were living in a very different universe from the one commonly assumed.

A revolution had taken place in cosmology, and I was proud that we had played a role in it. We had thrown out the old notion that galaxies are distributed fairly evenly throughout the universe and had replaced it with a very different view. Some regions of the universe are virtually devoid of galaxies, existing as vast stretches of nothingness; in others, billions upon billions of galaxies are aggregated together in immense galactic superclusters, which exert enormous gravitational influence on galaxies many hundreds of millions of light-years distant. Our own Milky Way, as we discovered, is one such galactic victim, and is being dragged at six hundred kilometers a second toward a great, unseen, previously unsuspected supercluster.

This description of the universe—of huge galactic concentrations intermixed with unimaginable void—is a very different universe from the one accepted by astronomers at the beginning of the 1970's. This new view made it all the more urgent to understand the mechanism that formed cosmic structures after the big bang. The massive galactic conglomerations of today's universe must have grown from cosmic seeds present in the earliest times of the universe. These seeds should be evident as fluctuations in the cosmic background radiation—fluctuations that represent primordial regions of slightly higher density. These wrinkles in space-time would have triggered the local condensation of matter under the influence of gravity, producing the embryos of galaxies and superclusters. And yet no such seeds had been seen. The cosmic background radiation, as far as we or anyone else had been able to determine, was completely smooth in all directions. This must mean one of two things. Either cosmologists' theories were badly wrong. Or no one had looked hard enough for the seeds.

I was convinced at the time that big bang theory was valid, and that galaxies and galactic structures were formed by gravitational collapse upon cosmic seeds. I therefore made up my mind to look even harder, to stretch technical capabilities—and human capabilities—to the limit. Modern cosmology rested on four pillars: the dark night sky; the composition of the elements; the expanding universe; the existence of the cosmic background radiation. The wrinkles—if we found them—would provide a fifth pillar. Their discovery would therefore be a major event in modern cosmology.

In the late 1970's, I thought the search wouldn't take very long—five years, perhaps ten at most.

Chapter 8

The Heart of Darkness

Vera Rubin had tried, and failed, to persuade astronomers that many galaxies display unusual motion or peculiar velocities superimposed on the general Hubble expansion of the universe. Back in December 1950, when the twenty-two-year-old Rubin presented her data to a meeting of the American Astronomical Society at Haverford College, Pennsylvania, few were prepared to listen. Her message was so contrary to established wisdom that the audience was effectively deaf to her data and her arguments. Rubin subsequently stopped studying peculiar velocities; she spent the 1950's and 1960's raising her children, completing her doctorate at Georgetown University, and working with Geoffrey and Margaret Burbidge at the University of California at San Diego. Previously she had relied on galactic redshifts determined by other astronomers; with the Burbidges she learned how to gather her own redshifts using a large telescope.

Working with Kent Ford at Kitt Peak National Observatory in Arizona, not far from where Slipher did his famous measurements of the redshifts of galaxies, Rubin embarked on a series of observations that would undermine orthodox views of the universe. This time, however, Rubin's remarkable discovery was quickly embraced, and it has emerged as part of cosmology's most tantalizing puzzle.

In November 1977, Rubin and Ford began testing a new technique for studying the motion of stars within spiral galaxies, of which our own Milky Way is an example. Spiral galaxies appear as flat disks where the majority of stars are concentrated at an extremely bright center and are rarer toward the dim periphery. The stars in the trailing arms orbit around the massive center all in the same direction, together composing a rotating galaxy. This is analogous with the Solar System, in which the planets orbit the same way around our massive sun, under its gravitational influence.

Ford had developed an image intensifier with which he and Rubin recorded on photographic plates spectra across a whole galaxy—from the brightest stars near the center of the galaxy to the faintest at the edges of the spiral arms. They were able to determine the velocities of stars at all distances from the center. This was a first in astronomy. Newton's law of gravity makes a simple but firm prediction about each star's velocity, depending on its distance from the center of the galaxy.

One of the great triumphs in cosmology was Newton's extrapolation of the law of terrestrial motion and gravity (the falling apple) to the Moon and planets. According to Newton, planets at the periphery of the Solar System (like Pluto) orbit at speeds much lower than planets close to the central massive Sun (like Mercury). This prediction was confirmed by observation—Pluto orbits at about ten thousand miles an hour and Mercury orbits at ten times that velocity. The same should be true for spiral galaxies: The more distant a star is from the center, the lower its orbiting velocity.

Excited about trying their new technique the first night, Rubin left Ford operating his intensifier and the telescope while she rushed downstairs to the darkroom to develop the first photograph. Rubin was startled at what she saw: The predicted relationship between the velocity of a star and its distance from the center of the galaxy did not hold up. The results showed that even stars in the galaxy's far periphery orbited at nearly the same speed as those closer to the center. There were only two possible explanations: Either Newton's law of gravitation fails at galactic scales, which would have shaken physics to its foundations; or galaxies are not what they seem.

To our eye, galactic mass appears concentrated toward the center and diminishes toward the periphery. And yet the stars at the periphery move as if they are embedded in much greater mass; so much, in fact, that this unseen mass must extend far beyond the periphery. If this inference is correct, then galaxies are truly *not* what they seem; the visible part—the stars we see—must be swamped by an immense quantity of invisible mass. Faced with such a dramatic conclusion, Rubin and Ford pressed on urgently with more measurements. In each case they examined, the same pattern emerged: Stars at the periphery of spiral galaxies move too fast—*if* one assumes the luminous part of these galaxies represents their entire mass. With data collected on ten such galaxies, Rubin and Ford were ready to publish, which they did in 1978. The unseen matter that, by inference, must be a major component of galaxies came to be known as "dark matter." Astronomers

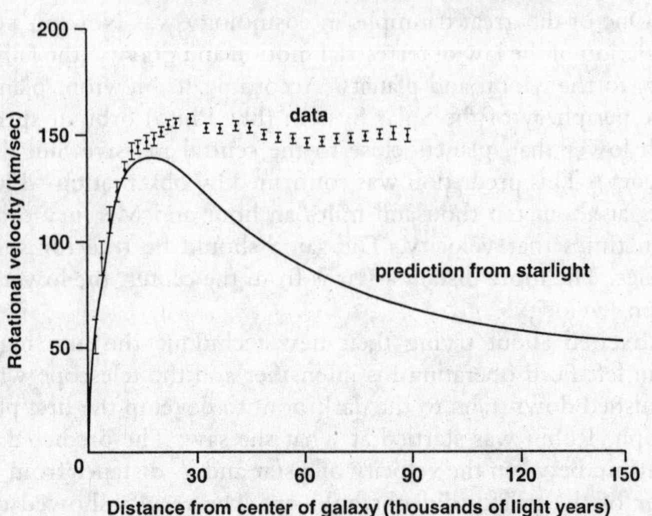

The bars show the measured velocities of stars around a spiral galaxy. The line shows the expected velocities of stars around the galaxy, assuming starlight reveals all the matter in the galaxy. The discrepancy indicates that there is dark matter in the universe—and much more of it than luminous matter. CHRISTOPHER SLYE had to come to terms with the fact that for generations what they had seen through their telescopes was a fraction of the stuff of the universe—the stars that illuminate our night sky may be only a small part of the products of creation.

The high quality of Rubin and Ford's data undoubtedly was an important factor in the rapid acceptance of the reality of dark matter, but there were other reasons. For years there had been scattered suggestions from observation that spiral galaxies are surrounded by dark halos, cloaks of unseen matter.* However, such a notion was so far outside conventional theory that it wasn't taken very seriously. The Swiss astronomer Fritz Zwicky had argued since the 1930's for the existence of unseen matter, based on the unusual motion of galaxies. Then, in 1974, Jeremiah Ostriker, James Peebles, and Amos Yahi at Princeton University predicted that something like dark matter must exist. Their calculations on the gravitational stability of spiral galaxies implied that such structures would fragment as they rotated, because of vibrations triggered by their uneven composition. If, however, the visible disk was embedded in a much larger and extended (unseen) mass, then such vibrations would be damped, and the spiral would remain stable. The very existence of our Milky Way implies the reality of unseen matter, they surmised.

Sometimes in science, different threads of evidence of the same phenomenon intertwine to create a compelling argument. Such was the case with dark matter. Within a couple of years of the publication of Rubin and Ford's paper, dark matter had become an obsession among cosmologists. The immediate questions were these: How much dark matter is there? And what is it? There are deeper questions, too, ones that are fundamental to our understanding of the universe: What role did dark matter play in the formation of galaxies? What does dark matter imply for the origin of the universe? And what of the fate of the universe? These latter questions inevitably interweave with the major issues in this book, and particularly with the search for the primordial seeds from which our present universe developed.

*Astronomers now think galactic halos are up to thirty times the diameter of each visible galaxy. If these halos were visible to the naked eye, then our sky would be even more spectacular than it is; from Earth, we'd be able to see more than a thousand galaxies with halos wider than the full Moon.

* * *

What happens to the universe is determined by its contents—as Mach and Aristotle would have wanted. If there is sufficient mass in the universe, gravitational forces will be strong enough one day to bring the post–big bang expansion to a halt and even reverse it, leading to a cataclysmic "big crunch." If, however, there is insufficient mass to cause this to happen, expansion will continue forever, with the temperature of the universe steadily falling. This is often referred to as the "big chill." Whichever fate awaits us—and we can't yet be certain which it is—it is comfortably distant, at least 50 billion years in the future. There is, however, an escape from these uncomfortable prospects. If the density of mass in the universe is poised precisely at the boundary between the diverging paths to ultimate collapse and indefinite expansion, then the Hubble expansion may be slowed, perhaps coasting to a halt, but never reversed. This happy state of matter is termed the critical density.

The critical density is calculated to be about five millionths of a trillionth of a trillionth (5×10^{-30}) of a gram of matter per cubic centimeter of space, or equivalent to about one hydrogen atom in every cubic meter—a few in a typical room.* This sounds vanishingly small, and is. It is an average figure and does not give a sense of the enormity of extremes that is our universe, ranging from unbelievably high densities in some regions to complete voids in others. If we know the critical density, then we can—in theory—begin to figure out our fate. All we have to do is count up all mass in the universe and compare it to the critical density. The ratio of the actual density of mass in the universe to the critical density is known, ominously, by the last letter of the Greek alphabet, Omega, Ω. An Omega of less than 1 leads to an open universe (the big chill), and more than 1 to a closed universe (the big crunch). An Omega of exactly 1 produces a flat universe.

The terms "flat," "open," and "closed" refer to the curvature of space. In 1915, Einstein introduced his general theory of relativity, according to which gravity is caused by the curvature of space-time. Einstein believed the universe is so massive that it

*This value is not precise since we do not know the Hubble expansion rate very well.

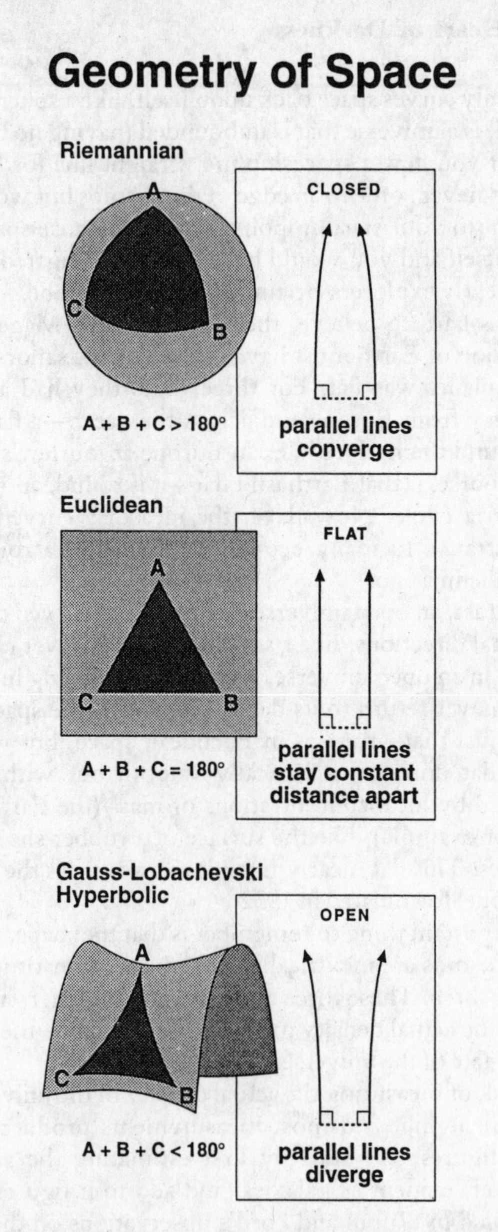

How is space shaped? That depends on how much matter there is. Space is Riemannian, Euclidian, or hyperbolic—closed, flat, or open, respectively—depending on whether its density is greater than, equal to, or less than the critical density. CHRISTOPHER SLYE gravitationally curves space back upon itself, like a sphere. The result, he said, is a universe that is unbounded (having no boundary) but finite: If you flew a spaceship in a straight line for billions of years, you'd never come to an edge of the cosmos but would eventually return to your starting point. That is because space curves back upon itself and you would have "circumnavigated" the cosmos, as the early explorers circumnavigated the globe.

If this is hard to believe, then imagine how Magellan's circumnavigation of Earth must have astounded his sailors, who assumed the planet was flat. For three years they had apparently traveled away from Europe and across the ocean—a flat surface. Yet they found themselves back in Europe at journey's end. The reason, of course, is that Earth isn't flat—it is round, and their ship had gone in a circle. Nowadays, the idea of a curved universe sounds as strange to many people as the idea of a round Earth sounded millennia ago.

In contrast, an open universe is curved, but curved oppositely in orthogonal directions, like a saddle, so that it never closes back upon itself. In an open universe, a spaceship traveling in a straight line would never return to its departure point. The space in a flat universe is just that—flat—as in Euclidean space, but with small ripples. A flat universe is basically smooth but with scattered warps caused by local concentrations of mass (the Earth, Moon, and Sun, for example), like the surface of a rubber sheet covered with marbles. This flat, nearly Euclidean universe is the Einstein–de Sitter model formulated in 1932.

The important thing to remember is that the shape, mass, and fate of the cosmos are inextricably linked; they constitute a single subject, not three. These three aspects come together, in Omega, the ratio of the actual density to the critical density—the ratio that predicts the fate of the universe.

The task of measuring the actual density of the universe is extremely challenging, and most measurements produce only approximate figures. We start by first estimating the amount of visible matter, evident as galaxies, and add to it figures for dark matter implied by Rubin and Ford's observations on the velocity of stars in galaxies. Stellar mass and other visible matter in the universe amount to no more than 1 percent of the critical density. When one adds the estimated amount of dark matter in galaxies, the figure rises to 10 percent or less, giving an Omega no more than 0.1 (one tenth) and a chilly fate for the universe. However, there are more pockets of dark matter to be inferred. When astronomers measured the relative speeds of pairs of galaxies thought to be orbiting each other, they found that the dark halos of galaxies must extend out nearly as far as the nearby orbiting companion, which implies more dark matter. This approach was extended to clusters of galaxies and superclusters of galaxies, and in each case these celestial structures move as if embedded in larger and larger swaths of dark matter. Moreover, peculiar velocities of galaxies deep in space—measured by several teams of researchers in the United States, Canada, Britain, and Australia—require even more dark matter. The galaxies are being attracted at huge velocities by unseen masses.

What's the bottom line? If we total all the potential pools of dark matter, and throw in the insignificant amount of visible matter, then we arrive at an average density of the universe of close to the critical density: Omega is close to 1. (There is, however, still some uncertainty and argument over these results.) In any case, it appears that dark matter may constitute as much as 99 percent of the stuff of the universe—a truly sobering thought.

In the 1970's Robert Dicke and Jim Peebles went around giving talks explaining why Omega *must* be near 1. Often they called the talks "The Flatness Problem." Their argument went like this: We know that the stars alone mean that Omega is greater than about 0.01, and the fact that the universe is not collapsing means Omega is less than 2. Since the density is near the critical density now, then it must have been closer in the past. The further back ones goes toward the big bang, the closer it gets to 1. If one goes back to about three minutes after the big bang, the time when Gamow and Weinberg tell us helium and the light elements were just finished cooking, then the density of the universe was about that of water, and Omega must be within one hundred million millionths (10^{-14}) of 1. At one second, Omega would be within one ten-millionth of a billionth (10^{-16}) of 1. And at the shortest time imaginable, the Planck time (10^{-43} seconds)—right at the birth of the universe—it would be within about 10^{-60} of 1. That is so close to 1 that reasonable people think it is not simply a matter of chance—something requires that Omega is 1 to all those decimal places.

A second theoretical reason why Omega must be nearly 1 is that it is easier to produce structure—that is, galaxies, clusters, superclusters, and compensating voids. During the 1980's, diligent astronomical observations revealed that the universe is much more highly structured than ever imagined. For instance, my colleague Marc Davis began a program to conduct a systematic survey of the sky, obtaining the redshifts of and distances to thousands of galaxies. It is now clear that galaxies are not scattered more or less randomly through space, as had once seemed the case, but are aggregated as clusters and superclusters, like a cosmic foam where the walls of the bubbles are concentrations of galaxies, while the bubbles' interiors are vast regions of empty space—much like a sponge. And, based on the peculiar velocities of our own and neighboring galaxies, a group of astronomers calling themselves "the Seven Samurai"* have inferred the existence of a cosmic behemoth—the Great Attractor, discussed in Chapter 7—some 150 million light-years away and nearly that large, in the direction of the Hydra-Centaurus supercluster. Populated by tens of thousands of galaxies, such a body would represent a truly unimaginably immense concentration of matter.

In 1986, a landmark paper was published by Valerie de Lapparent, Margaret Geller, and John Huchra. They found, in addition to this foamlike disposition of galaxies through space, evidence for even larger structures that are hundreds of millions of light-years in extent (for comparison, the Milky Way is a mere sixty thousand light-years across). One of these massive sheets of galaxies, named the Great Wall, is a kaleidoscope of thousands of galaxies. As a balance to these huge concentrations, immense voids also exist. A famous one is named the Great Void in Boötes and is of the order of 100 million light-years across.

As a result of this observational and theoretical evidence, it is clear that the current universe is much lumpier than we thought not so long ago. However, the great concentrations of mass we see in the clusters, superclusters, and higher-order structures are

*The Seven Samurai are David Burstein, Roger Davies, Alan Dressler, Sandra Faber, Donald Lynden-Bell, Roberto Terlevich, and Gary Wegner.

matched several times over by accompanying dark matter. Just how much dark matter is present within these structures—as opposed to scattered between them—determines how lumpy the universe is.

From our U-2 observations of the cosmic background radiation and other measurements, we already knew the primordial variations—the wrinkles—in the early universe were small. The wrinkles were "seeds" where matter gravitationally accreted, forming today's structures. Because the wrinkles were so small, there are limits on how much matter must be in the universe, if the structure we see today formed in the 15 billion years since the big bang. If the universe has an *average* density close to the critical density, some regions would have an effective Omega greater than 1, and others less. Those regions with an Omega of greater than 1 would eventually collapse and form structures, while less dense regions would become voids—leading to the foamlike structure of today's universe. If Omega were well below 1, however, then very few regions would collapse. If Omega were well above 1, then everything would collapse. The closer Omega is to 1, the easier it is to form the structure of the universe that astronomers now observe.

What is this dark matter that appears to constitute so much of the universe? The matter with which we are most familiar is assembled from protons and neutrons, collectively known as baryons, and the electrons that accompany them. Is the dark matter also baryonic in nature, but not shining and therefore invisible? There is no theoretical reason for ruling this out, and calculations of primordial nucleosynthesis (the making of nuclei from the protons and neutrons in the early universe) during the big bang and subsequent events allow us to estimate the density of baryonic matter—both visible and dark. A baryonic matter density 100 percent of the critical density would give us the comfort of knowing that all matter is composed of the same fundamental particles of which we and all things we see are composed. Any figure less than 100 percent would force us to acknowledge that the universe is made up of three kinds of matter: the baryonic visible matter, which we've long known about; the baryonic dark matter, which, though novel to us, is built of the same fundamental particles; and a third, completely unknown class.

According to calculations of nucleosynthesis in the big bang by Dave Schramm of the University of Chicago and co-workers, baryonic matter constitutes no more than 10 percent of the critical density. We are therefore forced to contemplate the fact that as much as 90 percent of the matter in the universe is both invisible and quite unknown—perhaps unknowable—to us. A mighty big mystery began unfolding, and the existence of dark matter—baryonic and nonbaryonic—is a major part of that mystery.

There is no shortage of candidates for the least exotic of our quarry: the baryonic dark matter. It may come in many forms—clouds of gas or dust, large planetlike objects, various forms of degraded stars, and black holes. True, some of these are among the more enigmatic bodies in the universe, but at least they are known to exist—and may exist in sufficient quantity to account for the entire complement of baryonic dark matter.

It is unlikely that dust or hydrogen and helium gas—either thinly distributed or as clouds—could be a major component of baryonic dark matter, because they would be readily detectable, principally by affecting the visibility of galaxies beyond them. However, if gas or dust condensed to form dense objects, such as small brown dwarf stars or large planetlike objects, then large amounts of dark matter might be effectively hidden from immediate view. My colleagues Kim Greist, a theorist, and Bernard Sadoulet, a French physicist who heads our Center for Particle Astrophysics at Berkeley, call bodies of this sort massive compact halo objects, or MACHOs. In addition to brown dwarfs and planetlike objects, MACHOs could include black holes and burned-out stars, such as white dwarfs or neutron stars, as proposed by my Berkeley colleague Joseph Silk.

Black holes are perhaps the most intriguing, and the most difficult to detect and quantify. As far back as the eighteenth century, scientists speculated about worlds so massive that nothing escaped their gravitational grip, not even light. In the early twentieth century, J. Robert Oppenheimer used Einstein's general theory of relativity to explain how a black hole might form: The black hole would warp adjacent space so deeply that the escape velocity would exceed the speed of light, which nothing can surpass; hence nothing, including light, could leave a black hole.

Black holes may inhabit the centers of galaxies, including ours. The center of the Milky Way emits intense gamma radiation—the "death cry," perhaps, of stars falling into a black hole. Black holes may also be distributed in galactic halos, where they might constitute a substantial fraction of baryonic dark matter.

Detection of baryonic dark matter is, in principle, feasible, because it is composed of the same fundamental particles with which we are familiar. But it will require some ingenuity. If most baryonic dark matter is composed of some form of MACHO, then it is highly concentrated and may have to be detected by its effect on other, visible objects, not directly. For instance, Bohdan Paczyński, a Princeton astronomer, suggested a way to do so almost a decade ago. When a MACHO passes in front of a distant star, its gravity would slightly bend and focus the star's light, making it briefly brighter. Charles Alcock, of Lawrence Livermore Lab, is the head of such a search by our Berkeley Center for Particle Astrophysics and an Australian group. The search is focused on the Large Magellanic Cloud, a small galaxy some 160,000 light-years distant from the Milky Way. If the MACHO idea has any validity, there must be millions of such bodies in the Milky Way's halo, and each of them has the potential to briefly brighten a star. Of course, there are more than a million stars in the Large Magellanic Cloud, so the star watch will require persistence and patience.

This and other searches, remember, have as their quarry a mere 10 percent of the likely dark matter in the universe. The other 90 percent is nonbaryonic, being composed of particles other than protons, neutrons, and associated electrons. If the search for baryonic dark matter is considered challenging, the hunt for the nonbaryonic variety is vastly so. We have to remember that conditions in the early instants of the big bang were truly extraordinary, so that all kinds of matter could have been produced. Some of these exotic forms of matter may have been stable enough, abundant enough, and massive enough to become the dominant but invisible component of the present universe. Although this exotic form of mass would exert a gravitational influence on visible matter—hence the observed peculiar velocities of galaxies—it fails to interact electromagnetically, and is therefore invisible. What is it?

Astrophysicists have contemplated two main forms of nonbaryonic particle dark matter, hot and cold, each of which has had its moment as most favored candidate. The adjectives *hot* and *cold* refer to the expected velocities of the particles: Hot particles are fast-moving and dispersive, while cold ones are slow-moving and thus cohesive. (Baryonic dark matter may be thought of as a form of cold dark matter, because it, too, is slow-moving.)

One of the most promising candidates for hot dark matter is the neutrino, of which three forms exist—the electron neutrino, muon neutrino, and tau neutrino. These subatomic particles are everywhere; billions of them zip harmlessly through your body while you read this sentence. Until the late 1970's, neutrinos were thought to be massless particles (like photons). Then U.S. and Soviet scientists independently reported evidence that electron neutrinos might have a slight mass of about thirty electron volts,* which is close to 0.0000001 percent of the mass of a hydrogen atom. If true, the total mass of neutrinos in the cosmos could account for the projected quantity of nonbaryonic dark matter, and might even be sufficient to close the universe (that is, to trigger eventual collapse). It was later shown by experiment and by observations of neutrinos from supernova 1987A that the electron neutrino mass is significantly lower than thirty electron volts. However, some particle theorists speculate that the tau neutrino may have a mass that large.

There is a general objection to any form of hot dark matter. This relates to the particles' velocity and the requirements for beginning the process of condensing galaxies from ordinary matter. We know that galaxies formed relatively early in the history of the universe, perhaps as early as half a billion years after the big bang. The process of condensation from primordial matter must have begun soon after the separation of matter from radiation, some ten thousand to one hundred thousand years after the big bang. This presents a problem for hot dark matter, because galaxies would have formed much later than we know they did.

Particles of hot dark matter would have been moving at close to the speed of light since shortly after the big bang. As a result,

*At the atomic scale, it's convenient to refer to mass in electron volts, a unit of energy. This reflects Einstein's dictum that mass and energy are opposite sides of the same coin.

by the time the condensation process could have begun, any such primordial concentrations of matter would have been smeared out as the particles streamed out to huge distances. It turns out that the smallest structures that could be built under the influence of hot dark matter would be about 10 million light-years in size. This is the scale of galactic superclusters, not individual galaxies. From such superclusters, individual galaxies would later have condensed. With this so-called top-down dynamic, galaxies would have formed much later in the history of the universe than we know to be the case. Hot dark matter therefore has to be eliminated as a candidate for nonbaryonic dark matter, because it cannot have formed galaxies early enough. Unless, of course, something else provided stable galaxy-sized structures with which hot dark matter interacted.

One such possibility has been proposed: cosmic strings.

It sounds like a science-fiction movie: An immensely long, string-shaped object with awesome gravitational pull slices through the Solar System. Like a cosmic bullwhip, it tears through Earth and rips it apart, then disappears at almost the speed of light, leaving behind a lifeless world. Bizarre as it sounds, the existence of such objects has been championed. That they were proposed at all perhaps suggests how desperate cosmologists were to explain how structure—particularly large-scale structure—condensed from the early universe.

Cosmic strings are one example of a topological defect in the cosmos, and certainly the most famous. The concept of topological defects concerns symmetry, and how it is sometimes broken. Water is a good example: If you are submerged, water looks exactly the same in all directions; it has no structure. When water freezes into ice, it has a different symmetry. Ice consists of long crystal lattices of atoms that are oriented in specific directions. The water's degree of symmetry is related to its temperature, and at certain temperatures water undergoes "phase transitions"—say, when it turns from liquid into ice.

Something similar happened after the big bang. At the birth of our cosmos, the primordial universe was extremely hot and dense. The forces were united in a single, symmetric force—a primal force. The cosmos expanded and cooled. As the temperature fell, the original symmetry was broken and the primal force fragmented into subforces, each of which manifested itself in a different way. These forces are now differentiated into electromagnetism and the strong and weak nuclear forces. In other words, the symmetry-breaking process that turns water into ice cubes is akin to the process that spawned the universe around us. The key ingredient is temperature.

The big bang universe expanded to enormous size, cooling as it grew. At a certain point it was cool enough for symmetry breaking to occur and for the forces to differentiate—to freeze out in countless locations across space. Starting at each of these locations, the symmetry breaking fanned out across the universe at the speed of light. Inevitably, these ever-expanding regions of broken symmetry collided with each other. And when that happened, each expanding region's broken symmetry did not perfectly match up with the other regions it hit. This had dramatic consequences.

The most evocative analogy is that of a pond on a cold day. Different parts of the pond begin crystallizing, and as the ice spreads across the pond, different freezing regions encounter other freezing regions. However, one region's crystals rarely link perfectly with those of other regions. And that's why ice in a pond—or in an ice cube, for that matter—is crisscrossed with thin white lines, which are really fractures marking where the crystals align improperly.*

In the cooling universe, symmetry breaking could have caused different regions to align improperly. As a result, flaws in space formed. These flaws retained the original symmetric superhot, supermassive state of the big bang. The flaws could have manifested themselves in various ways, including magnetic monopoles, domain walls, textures, or cosmic strings. Magnetic monopoles would be "zero-dimensional" defects—in effect, "points," without height, width, or depth. Domain walls would be two-dimensional, like immense sheets spread across space. Textures would be three-dimensional. And cosmic strings would be one-dimensional objects, with length but no width or breadth.

*The same thing is seen in a flawed diamond: Certain parts of the crystal lattice aren't perfectly aligned. Some flaws are stress fractures.

Bristling with primal energy, strings would be terrifying indeed.

Originally, cosmic strings were speculated about by researchers including Alexander Vilenkin of Tufts University, Thomas W. B. Kibble of Imperial College in England, and the Russian theorist Yakov Zel'dovich. But their most vocal champion was Neil Turok, a tall, thin young Englishman who wears black Clark Kent glasses and maintains a half-amused aplomb in the face of critics. A roving scholar who has worked at Princeton and other schools, he and his colleagues used computers to simulate the evolution of strings after the big bang. As massive as galaxies, yet as thin as threads (and perhaps as long as the width of the cosmos), strings would flick across the universe, perhaps occasionally splitting off loops as a string crossed over itself. These cosmic strings could have been the catalyst that helped the hot dark matter begin early galaxy formation.

It was an exciting idea, and the media loved it, but it eventually fell out of favor. One reason was that computer simulations showed strings would have been too unstable and weak to form galactic structures. "The better the computer simulations became, the more the virtues of cosmic strings began to disappear," Turok, waving a figurative white flag, announced. "Anyone who wants my string codes can have them." By the late 1980's most cosmologists had shifted their interest in nonbaryonic dark matter from hot to cold. A few, like Turok and his colleagues David Spergel of Princeton and David Bennett of Livermore, continue flying the defect flag in an attempt to keep their colleagues honest. In the early 1990's they argued that another defect—textures—could be a viable alternative, but they made little headway in convincing their colleagues to favor defects.

Nonbaryonic cold dark matter might seem like the dull cousin of the hot variety, lumbering about the cosmos at a mere fraction of the speed of light, but it does fulfill many of the requirements for the early formation of galaxies. Given their low velocity and low rate of interaction with light, particles of cold dark matter could have begun aggregating under the influence of the primordial ripples; they would quickly form galaxy-sized seeds that do not smear out. Baryonic matter would then accrete upon these seeds, forming galaxies within a billion years after the big bang. Galactic clusters and superclusters would form later, giving a bottom-up scenario to the creation of the overall structure of the universe. If this idea is sound, then we inhabit a baryonic solar system spawned from nonbaryonic seeds. The exploration of nonbaryonic dark matter could transform astronomy, the astrophysicist Scott Tremaine has written. He suggests we are experiencing "the first stage of a revolution against 'baryocentric' cosmology that is the direct descendant of the revolution that Copernicus led against geocentric cosmology."

Computer simulation gave a considerable boost to cold dark matter's claim to be the majority stuff of the universe. For instance, models of galaxy formation produced by Joel Primack and George Blumenthal of the University of California at Santa Cruz, and based on the properties of baryonic matter and nonbaryonic dark matter, closely mimicked the process as it is understood in the cosmos. And my Berkeley colleague Marc Davis showed that the large-scale structure of the universe emerged more reliably from computer models based on cold dark matter than they did from hot. Canadian theorist Dick Bond and his English colleague George Efstathiou pushed the theory further.

Although the properties of cold dark matter appeared consonant with the history of the universe as we understand it, the question of its identity was a real challenge. In response, cosmologists have come up with a family of hypothetical particles, known collectively as weakly interacting massive particles, or WIMPs, a name coined by Michael Turner of the University of Chicago. The offspring of both grand unified and supersymmetric theories, WIMPs must be long-lived and stable, must all have the same mass, and must interact only feebly with baryonic matter. Like Dirac's antimatter concept of the 1920's, supersymmetry theory holds that for every ordinary particle, there is a mirror, supersymmetric particle. For example, every fermion (particles such as quarks and leptons) has a counterpart in the form of a boson (particles such as photons and gluons). None of these mirror particles has been detected yet, but that hasn't discouraged physicists from assigning them weird names. For the photon there is, in theory, a supersymmetric particle dubbed a "photino"; for the quark, a "squark"; for the neutrino, a "sneutrino"; and so on.

There are other nonbaryonic dark matter candidates beyond

WIMPs, such as a heavy neutrino or a lightweight particle called an axion. Weighing in at twice the mass of a proton, the heavy neutrino would not be a true neutrino at all, and in any case would require a revision of the standard model of particle physics for it to be able to exist at all. Building on work by Robert Pecci and Helen Quinn, Steven Weinberg and Frank Wilczek predicted the existence of a particle that Wilczek dubbed the axion. The axion could be a dark matter candidate if it displayed a diminutive mass of less than a billionth of an electron's mass and had a virtually infinite lifetime. According to theory, these hypothetical particles could have been produced in vast numbers at the time when protons and neutrons were forming from the aggregation of quarks and might even have been the dominant form of matter at that time.

The list of nonbaryonic dark matter continues further, and ever more whimsically. Some such suggestions include so-called quark nuggets and little black holes, both of which require rather unusual initial conditions. Are such putative forms of matter the fantasies of desperate men and women, frantically seeking solutions to baffling problems? Or are they a legitimate sign that with the discovery of dark matter cosmology finds itself in a terra incognita beyond our immediate comprehension?

Whatever is the case, several research teams have recently set up experiments to detect the presence of WIMPs and other nonbaryonic dark matter particles, which, given how little is known about this shadow world, is a considerable technical challenge. For instance, physicists at Livermore National Laboratory are trying to detect axions by using a copper cylinder surrounded by a superconducting magnet; as axions passed through the magnetic field, they would make the cylinder vibrate slightly and emit microwaves. Other groups plan to detect WIMPs in cosmic rays. For example, Bernard Sadoulet and his colleagues hope to measure vibrations and ionization generated by WIMPs shooting through a supercooled detector. Their device will be tested at Stanford in a hole more than sixty feet deep that provides shielding against cosmic rays and other interference.

Also at Stanford, physicist Blas Cabrera is developing a detector that contains superconductors; these would lose their superconductivity when slightly heated by a passing WIMP. In

England, Peter Smith of the Rutherford-Appleton Laboratory aims to place WIMP detectors in a mine more than three thousand feet deep on the coast of northeastern England. One of his detectors contains a crystal of sodium iodide; it will emit a flash of light or scintillate if its nuclei or electrons are clobbered by a WIMP. In Italy, Ettore Fiorini of the University of Milan plans to install WIMP detectors in the Gran Sasso Tunnel in the Apennine Mountains.

As we wait to hear of the first sighting of one of these exotic particles, there is a lingering worry. The theoretical strength of nonbaryonic cold dark matter—its facility to clump together in the early formation of galaxies—may also be its weakness. The effective clumping on the galactic scale would mean less dark matter on larger scales, that is, beyond the galaxies and their dark halos. Cold dark matter alone might not be sufficient to explain the total amount of clumping on larger scales—clusters of clusters of galaxies. Something else would have to be added—which means turning back to add some hot dark matter. And so it goes.

In 1977 when Vera Rubin and Kent Ford made the discovery that effectively established dark matter as a legitimate cosmological puzzle, Rubin thought the puzzle would be solved in little more than a decade. The decade has more than passed, and we are no further forward in meeting the challenge. As Rubin says, "I really thought we would know more by now. It is really disappointing to realize that the answer is as far away as ever."

During the search, however, our appreciation of the universe and its origins has sharpened considerably. We faced a major paradox: With its foamlike structure of galaxies, its clusters, superclusters, and other massive structures, the present universe is evidently extremely clumpy; and yet, we know from our measurements of the cosmic background radiation that the distribution of matter in the early universe was virtually uniform. The challenge was to explain the early pervasive homogeneity, which somehow, with dark matter in whatever guise, evolved into the present staggering heterogeneity.

The potential answer wasn't long in coming: inflation.

Chapter 9

The Inflationary Universe

On February 14, 1982, Stanford University physicist Blas Cabrera received an unusual Valentine's Day gift: a blip on a detector in his basement laboratory. The device was intended to detect particles called magnetic monopoles (isolated magnetic north or south poles), which are expected as relics of the big bang. Cabrera had recently switched on his wonderful new superconducting detector for the first time. Could it really have detected one of these elusive particles so quickly? News media got wind of the event. *The New York Times* published a front-page story. Cabrera refused to admit TV cameras for fear it "would have turned the lab into a zoo." Hesitating to identify the blip as a true monopole, he awaited the next one's arrival. Days passed, then weeks, then months. Harvard physicist Sheldon Glashow sent Cabrera a limerick:

> *Roses are red, violets are blue,*
> *Now's the time for monopole two.*

But monopole two did not show up. After a decade of waiting Cabrera has all but abandoned hope. Most physicists suspect the 1982 "monopole" was a technical glitch, perhaps even a student's prank. Dozens of other scientific teams around the world tried to detect monopoles in the 1980's; all failed. The media's fascination with monopoles turned to cynicism. In 1985, *The Wall Street Journal* dubbed monopole searches "one of the wildest goose chases in the history of science."

Yet much was at stake, which is why monopole searches continue to this day. Various hypotheses aim to explain all known forces via a single equation—an equation "you can wear on a T-shirt," jokes Leon Lederman. These hypotheses are known as grand unified theories, or GUTs, and they predict that monopoles should exist. So where are these elusive particles?

Ironically, the answer may have already been found—on the same northern California campus—three years before Cabrera's erroneous discovery. In December 1979, Alan Guth proposed a hypothesis that was a dramatic extension of big bang cosmology. Known as inflation theory, Guth's idea was of an ultrarapid, ultrabrief big bang within the big bang, one that occurred in the first instant of creation and that essentially set the conditions for the future evolution of the cosmos. Guth wasn't the only one to arrive at this radical solution. Similar ideas had occurred to Katsuhiko Sato, of Japan, and to Alexei Starobinsky (followed by Andre Linde) in the former USSR. They did not publicize their insights, but Guth did, and it shook the world of cosmology.

At a stroke, inflation theory appeared to solve a string of problems that cosmologists had long grappled with, and it surely merits recognition as a third major intellectual revolution in cosmology. Its predecessors were the intellectual fruits of, first, Galileo and Newton, who showed that terrestrial physics and celestial physics are the same; and second, of Einstein, whose general theory of relativity described the expanding universe. Inflation is important because it links two seemingly unrelated subjects—astrophysics (the science of the incredibly large) and quantum particle physics (the science of the incredibly small).

Inflation is an extremely powerful concept, and explains three major issues in cosmology. First, it explains the paradox of an extraordinarily uniform early universe, as revealed by the smoothness of the cosmic background radiation, that somehow evolved into the evident lumpiness of the present universe. Second, it explains the absence of magnetic monopoles and other putative relics (such as cosmic strings) of the early universe, the absence of rotation of the universe, the flatness of space, the homogeneity of space, and even why Einstein's cosmological constant wasn't completely bogus. Third, it explains why the universe is expanding. Moreover, the universe is, according to inflation theory, much, much larger than anyone ever suspected; hence our tiny corner of the cosmos is even tinier than we thought.

That is quite a list of achievements for a single theory—if valid.

When I was a graduate student at MIT, my research field was particle physics. My thesis adviser and research director was a very energetic, enthusiastic, friendly bear of a man, David H. Frisch. In 1968, as the time for my comprehensive examination approached, I asked Dave and his colleague Lou Osborne to give me practice oral exam questions. In the group meeting room where we often lunched were a table and chairs, a large blackboard where I worked out the answers to their questions, and a bookcase holding all the issues of the *Physical Review*—a primary physics journal. After a few sessions I was getting more fluid and comfortable with my answers, and they seemed to be running out of questions. Apparently trying to think of a question, Dave pointed over to the bookcase and said, "Do you see the collection of the *Physical Review* over there? Go over to it." I thought he was going to have me pick an issue and explain some article or result. Instead he said, "Note how each year the *Physical Review* grows in size. The distance between the front cover of the first January issue and the back cover of the last December issue is increasing rapidly. How long will it be until the annual covers are moving apart faster than the speed of light?"

He knew this would be a shocking and provocative question. How could the covers move apart faster than the speed of light? Well, I thought, it might hypothetically, but what about the practical issues? Then it was time to focus and answer the question—after all, I was trying to learn how to pass the oral exam. That meant answering the question to the satisfaction of the examiners. I realized that, besides being provocative, the question honored the MIT tradition set and led by Vicki Weisskopf of teaching students to estimate.

In physics problems many factors might be important, but generally only one or two dominate. Being able to estimate things quickly and accurately tells the physicist what can be neglected and what needs to be studied and treated carefully. Glancing at the bookcase, I drew on the blackboard a rough plot of the distance between annual covers versus year. The curve rose rapidly with advancing years. It displayed exponential growth, which is the compound interest—or inflation—curve. Each year was about a fixed factor larger than the previous year. The typical time for doubling was about six years. Hence it was easy to estimate when a year's worth of the *Physical Review* would be more than a light-year long.*

Frisch and Osborne pressed on, asking me to explain how the covers could move apart faster than the speed of light. Was there a conflict here with special relativity? Would something intervene to keep the covers from moving apart faster than light? These questions were not so hard to handle.

This puzzle parallels a cosmological query: Can galaxies move apart faster than light in an expanding universe? I could imagine the situation where markers were set up many years ahead for each issue of the journal, with each year in a row behind the previous. If I had enough researchers, each in the correct place and each writing one article, they could leisurely write one short article and slip it into the appropriate place in an issue. For the year where the annual covers were a light-year or more apart, it would appear that the annual covers were moving apart at a rate exceeding the speed of light even though no individual part had moved very significantly. In another doubling time, the annual covers would appear to be moving apart at more than twice the speed of light. From year to year the issue covers would appear to be moving apart at a rate proportional to their issue number (or date of publication). This is the Hubble law for the *Physical*

*The 1968 *Physical Review* volumes covered about four feet of shelf space. If it continued doubling in length every 6 years, in 286 years, or A.D. 2254, the covers would appear to be moving apart at the speed of light. That is, if all the volumes from that year were stacked together they would reach a light-year. In fact, in 1970 the *Physical Review* split into four separate sections, A, B, C, and D. It kept growing but not at quite the same rate. However, more journals came into being. Last time I checked, the total number of scientific articles being published was still on an exponential growth curve. I expect that recently there has been some slowing due to budget constraints.

Review. Something similar holds for the cosmos: Instead of physicists writing articles and putting them into the issues, the universe is generating space and putting it between the galaxies. No galaxy has to move; there just must be more space put between them.

"Okay, okay," Frisch and Osborne concurred. "It is possible. Could you do it without setting everything up so many years ahead so that the information could get across those many light-years?" No; without guide marks the researchers would not know where to put their contributions and would put them near their neighbors. Consequently, the *Physical Review* would wander all over the map—the random walk of a drunk who's lost his coordination skills and sense of direction.

This is precisely the same problem one would have in setting up the universe. It is easy to have parts of the universe moving apart at greater than the speed of light (without violating special relativity) if space is expanding. If space is expanding, then two parts separated by a distance greater than the speed of light divided by the expansion rate must move apart faster than the speed of light even though neither of them is moving or moving very fast relative to its local neighbors or space-time. What is impossible is keeping things synchronized and matched. It is exactly this synchronization problem that leads to defects or a highly mismatched and lumpy universe. The universe is not like that at all.

The central issue is the astonishing uniformity of the early universe, as revealed by the cosmic background radiation. As our observations and those of other research teams have demonstrated, the intensity of the cosmic background radiation is identical (to at least one part in ten thousand) from all directions in the cosmos. We are receiving radiation from thousands of different regions of the universe; regions that were not in contact with each other since an instant after the big bang. And yet, each and every one of them displays nearly identical temperatures. In the absence of a direct interaction between these disparate regions, how is such uniformity possible? How could it be established and maintained?

Traditionally, big bang cosmologists made the arbitrary assumption that the universe is smooth because the universe was smooth initially. This kind of assumption is unsatisfactory, not just because it is like playing God—"Let it be so!"—but also because it explains nothing and begs the question of why it was smooth in the first place. Astrophysicists, like most scientists, feel more comfortable if they can see a mechanism for making things as they are, without the need for arbitrary assumptions. That is what the inflation concept offered to cosmology. Ironically, inflation theory was originally conceived not to explain the uniformity of the cosmic background radiation but, in part, to explain the absence of a mysterious particle.

In 1978, the thirty-one-year-old Alan Guth was a research fellow at Cornell. His friend Henry Tye came to him and suggested they figure out how many magnetic monopoles were generated by the big bang. "At that time, it sounded to me to be an almost crazy thing to think about," Guth later said. "I had never worked on cosmology." Decades earlier, Paul Dirac, of antimatter fame, had predicted the existence of magnetic monopoles; he said they would be massive particles with a single (mono) magnetic charge (north or south pole).*

Guth forgot about Tye's suggestion until the following spring, when he heard Steven Weinberg lecture at Cornell. Weinberg is a well-known particle physics theorist whose book *The First Three Minutes,* about the big bang and the origin of matter, had just been published. In his speech, Weinberg talked about using grand unified theory calculations to explain the cosmic abundance of baryons, that is, ordinary matter such as protons and neutrons. Guth was impressed; perhaps cosmological problems weren't as intractable as he'd thought. So he joined Tye in thinking about monopoles, specifically about their theoretical abundance.

Tye and Guth concluded that monopoles, which would be produced by the symmetry breaking at the end of the GUTs era of the big bang (about 10^{-34} seconds), should be about as common

*We are used to magnets being dipoles, that is, having two poles: north and south. Particle accelerator physicists often build quadrupole magnets having four magnetic poles—north, south, north, south—in a symmetric configuration. (*Mono* is the Greek prefix for "one.")

The U-2 aircraft from NASA-Ames flying above the Golden Gate Bridge; fog and San Francisco in the background. The high-altitude U-2 jets were famous for reconnaissance during the Cold War. Now they are used for scientific research.

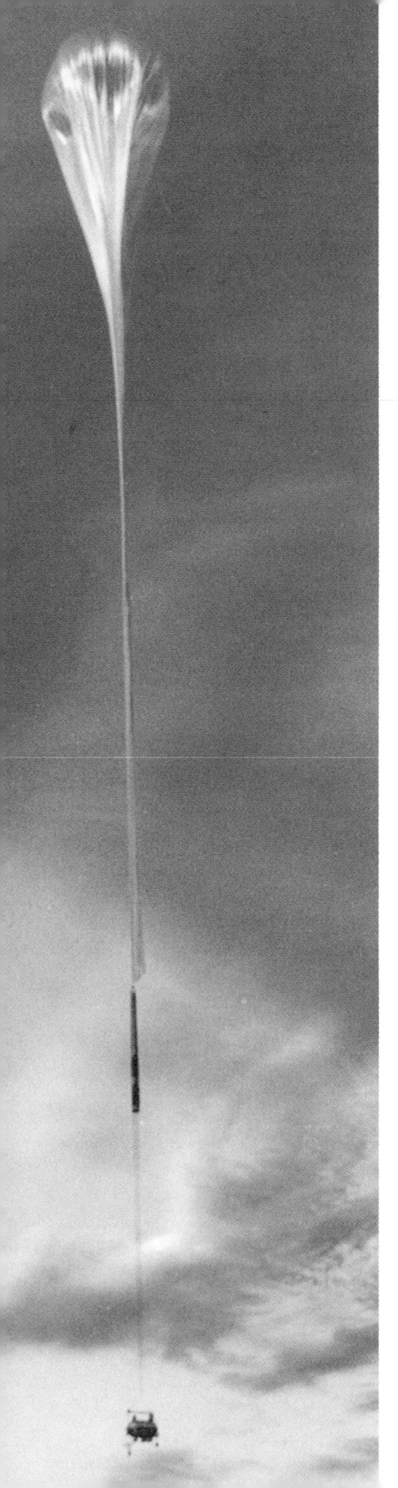

An antimatter experiment just after launch. The payload is about the size and weight of a heavy car. The parachute is one hundred feet long. When the balloon reaches a high altitude, the helium inside will inflate it to the size of a football stadium.

The active volcano Mount Erebus, Antarctica, photographed during the ski-plane flight from the coast to our experimental site at the South Pole.

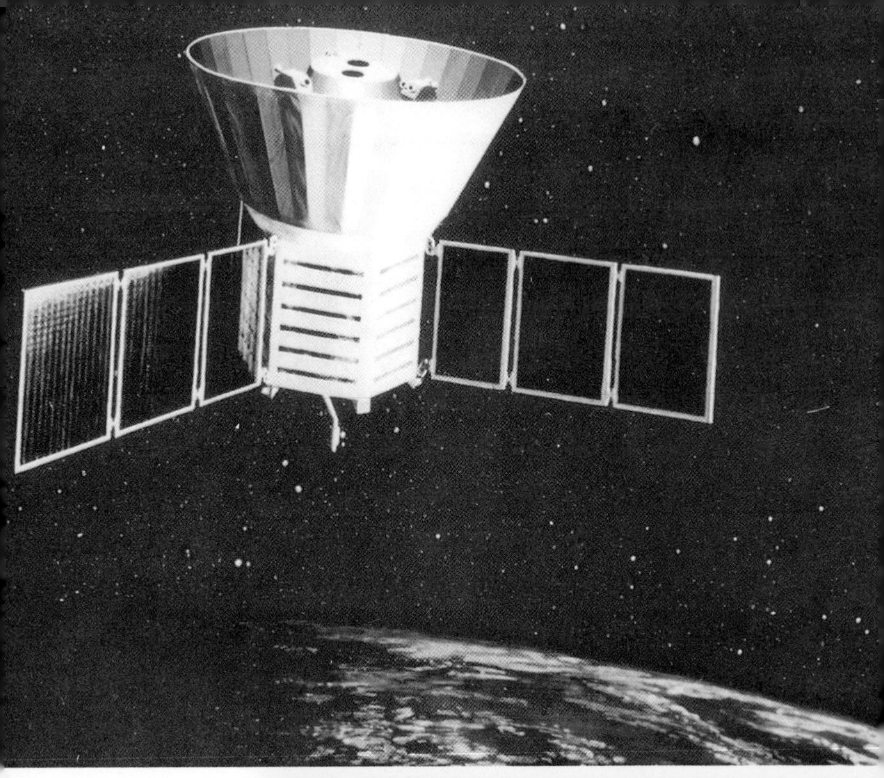

An artist's conception of COBE deployed in orbit.

The launch of the COBE satellite on a Delta rocket.

Three full-sky maps made by the COBE satellite DMR instrument show: top, *the dipole anisotropy caused by the earth's motion relative to the cosmic background radiation (hotter in the direction we are going, cooler in the direction we are leaving);* center, *the dipole-removed sky showing the emission from the plane of the galaxy—the horizontal red strip and the large-scale ripples in space-time;* bottom, *a map of the wrinkles in time.*

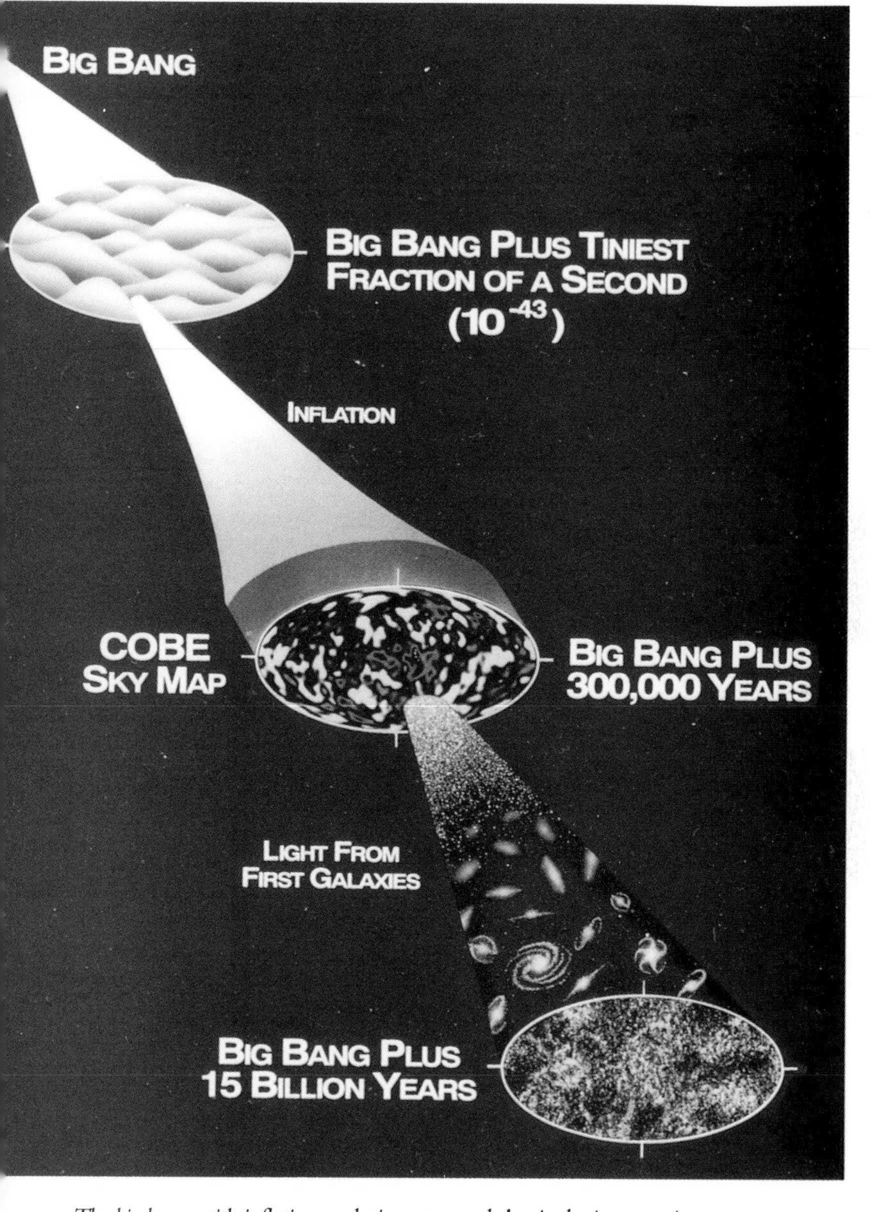

The big bang, with inflation producing space and the ripples in space-time mapped by COBE, and eventually evolving to become galaxies, clusters of galaxies, and larger structures.

1. *The origin of the universe—a pregeometry of foamlike space-time whose laws of physics are unknown.*

2. *At 10^{-43} seconds, temperature 10^{32} degrees Kelvin: The strong, weak, and eletromagnetic forces are unified into one indistinguishable force. This period is often referred to as the Grand Unification or GUT Epoch. During this epoch, there may have been a very rapid, accelerating expansion of the universe called cosmic inflation. This inflation made our universe very large and flat, but also produced ripples in the space-time it was creating.*

3. *At 10^{-34} seconds, temperature 10^{27} degrees Kelvin: The strong force becomes distinct from the weak and electromagnetic forces. The universe is a plasma of quarks, electrons, and other particles. Inflation ends and the expanding universe coasts, gradually slowing its expansion under the pull of gravity.*

4. *At 10^{-10} seconds, temperature 10^{15} degrees Kelvin: The electromagnetic and weak forces separate. An excess of one part in a billion (a thousand million) of matter over antimatter has developed. Quarks are able to merge to form protons and neutrons. Particles have acquired substance. This is the energy level that would be probed by the United States' SSC accelerator and CERN's Large Hadron Collider (LHC) accelerator.*

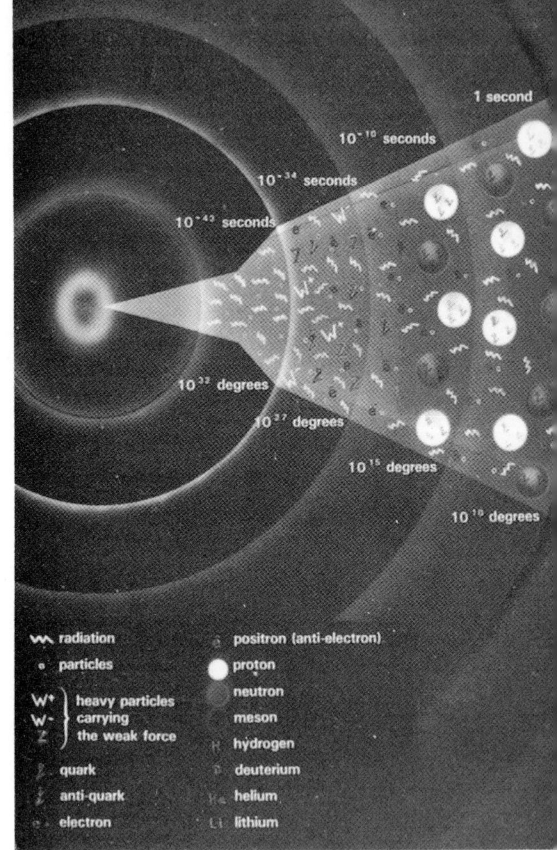

5. *At 1 second, temperature 10^{10} degrees Kelvin: Neutrinos decouple and then the electrons and positrons annihilate, leaving residual electrons but predominantly the cosmic background radiation as the main active constituent of the universe.*

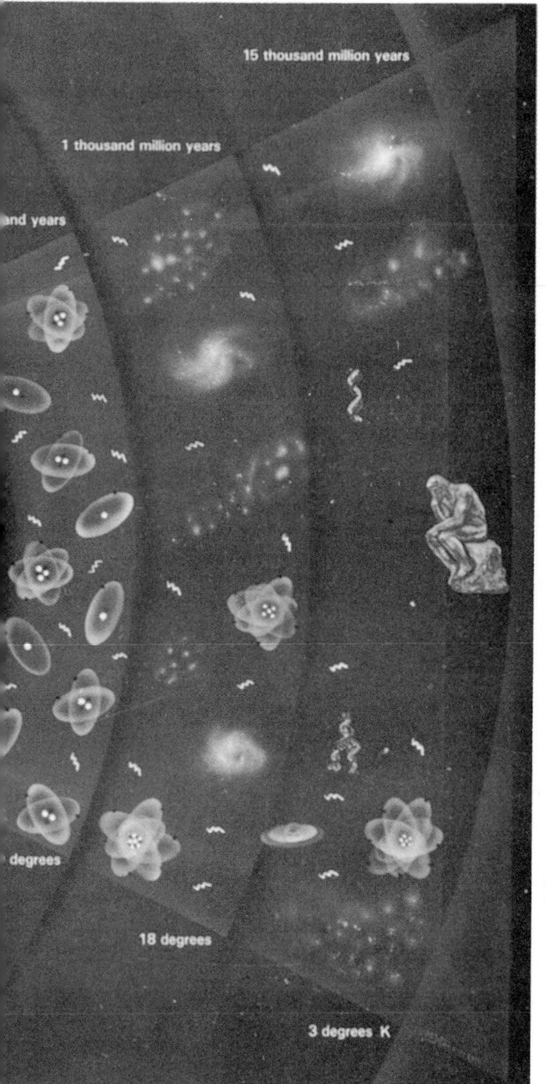

6. *At 3 minutes, temperature 10^9 degrees Kelvin: Protons and neutrons are able to bind together to form nuclei since their binding energy is now greater than the cosmic background radiation energy. A rapid synthesis of light nuclei occurs—first deuterium (heavy hydrogen: one proton and one neutron), then heavier elements, primarily helium but up to lithium (three protons and four neutrons) nuclei. About 75 percent of the nuclei are hydrogen (one proton) and 25 percent are helium (two protons and two neutrons); only a smattering are other elements. The heavier elements are later formed by nuclear burning in stars.*

7. *At 300,000 years, temperature 3,000 degrees Kelvin: Matter and the cosmic background radiation decouple as electrons bind with nuclei to make neutral atoms. The universe becomes transparent to the cosmic background radiation, making it possible for COBE to map this epoch of last scattering.*

8. *At 1 billion years, temperature 18 degrees Kelvin: Clusters of matter have formed from the primordial ripples to form quasars, stars, and protogalaxies. In the interior of stars, the burning of the primordial hydrogen and helium nuclei synthesizes heavier nuclei such as carbon, nitrogen, oxygen, and iron. These are dispersed by stellar winds and supernova explosions, making new stars, planets, and life possible.*

9. *At 15 billion years, temperature 3 degrees Kelvin: We have reached the present. Five billion years earlier, our solar system condensed from the remnants of earlier stars. Chemical processes have linked atoms together to form molecules and then complicated solids and liquids. Man has emerged from the dust of stars to contemplate the universe around him.*

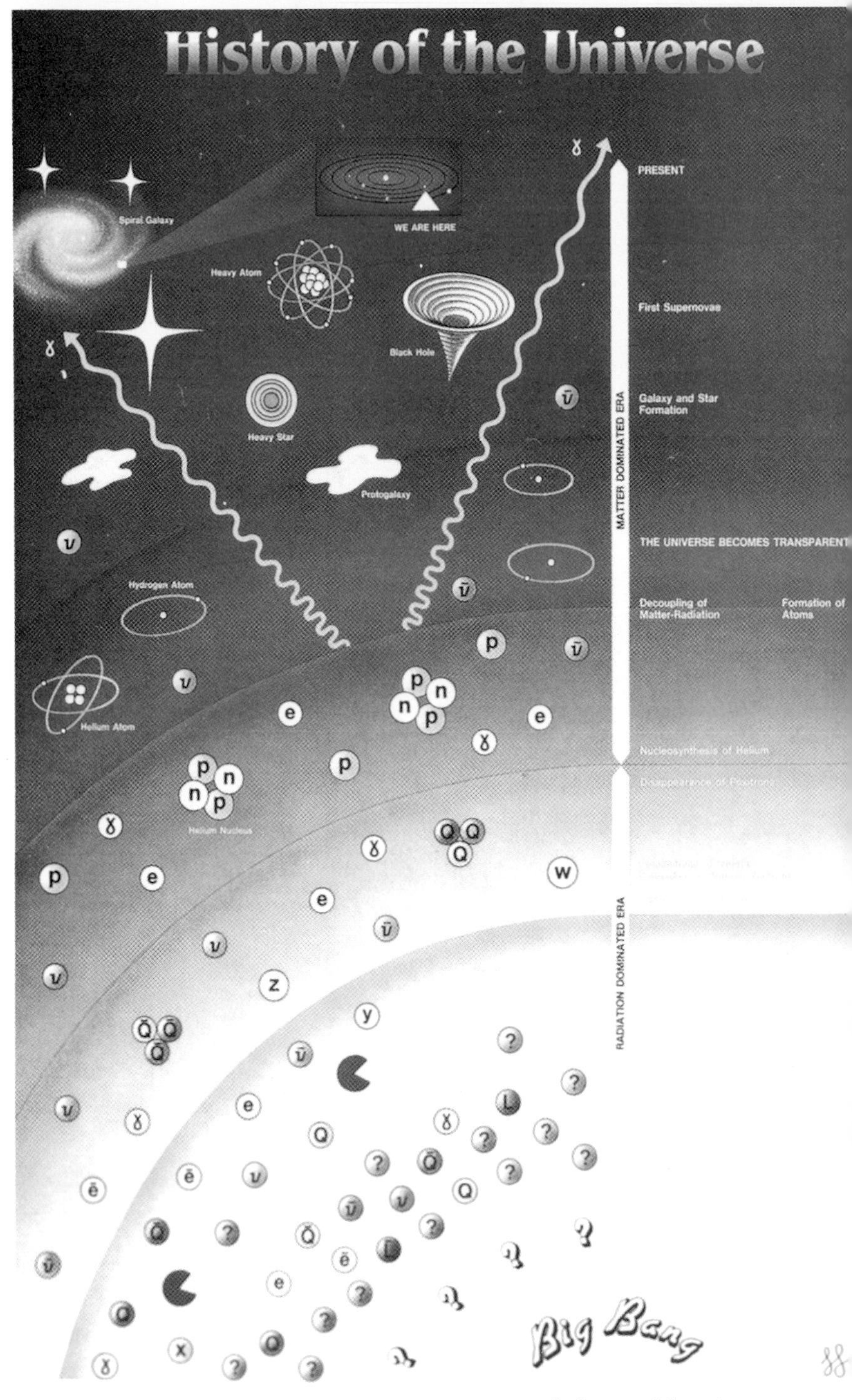

Time scale and important constituents in the history of the universe.

Alan Guth, pioneer inflation theorist. DONNA COVENEY, MIT

as gold atoms.* Gold is rare and precious, but you can buy it in any jewelry shop. So where were the monopoles? For years, researchers had searched in vain for these extremely heavy particles (a single one might weigh as much as a bacterium) in the oddest places: lunar rocks, Antarctic ice, even raw sewage. Perhaps monopoles are so elusive because they don't exist in the numbers predicted by the standard big bang model, Guth mused. Perhaps something else was needed?

*Symmetry breaking leads to topological defects—for GUT symmetry breaking at a time of 10^{-34} seconds, the simplest defects are magnetic monopoles. The same causality argument we had about synchronizing and smoothing would apply to the formation of defects. In 10^{-34} seconds light can travel only about 10^{-24} centimeters, so that there should be an average of one defect per region of that size. That small region has since expanded to be about three meters at present. Thus we would expect about one magnetic monopole per typical room.

The magnetic monopole was the first of two puzzles that steered Guth toward inflation theory. The second concerned the "flatness" of space. Late in 1978, Guth heard a speech by Princeton physicist Robert Dicke, a pioneer of big bang research, titled "Why Is the Universe So Flat?" Dicke pointed out that we know the space of our observable universe is very flat.

While Guth was contemplating the elusiveness of the monopoles and the phenomenal flatness of the universe, he hit upon the simple yet profound notion of inflation. If, in the first instant after the big bang, before it was 10^{-12} seconds old, the universe underwent an immensely rapid and accelerating burst of expansion, then these problems would be resolved. So, too, would the puzzle of a virtually uniform early universe; and other puzzles, too. Though many of these phenomena are interrelated, we'll pick our way through them one by one.

First, we will tackle the question of uniformity of the early universe, as revealed in the smoothness of the cosmic background radiation: How could such uniformity be established and maintained through the early life of the cosmos? The phenomenon of cosmic inflation does it with almost embarrassing ease.

If we take our minds back to the first instant of inflation, we find the answer. Back then, an eyeblink (say at 10^{-35} seconds) after the first moment of creation, all the potential mass and radiation of our part of the universe was subsumed in a primal soup of energy, parceled within a tiny region a trillionth the size of a proton (about 10^{-25} centimeters). Effectively, everything was connected to and equivalent to everything else—a primal homogeneity. Then our part of the universe experienced an incomprehensibly rapid eruption of space, so that by 10^{-32} seconds it had expanded to at least ten meters across. Next, inflation ended, and that ten-meter-sized region proceeded to expand at the more leisurely pace characteristic of the classical big bang to its present size, greater than one trillion light-years. In that tiny fraction of a second, our space expanded by a factor of more than one hundred times what it then did in the succeeding 15 billion years.

The homogeneity that existed in that tiny region at that early instant was spread across a region of the universe much larger than what we can see today. Inflation does not require the arbi-

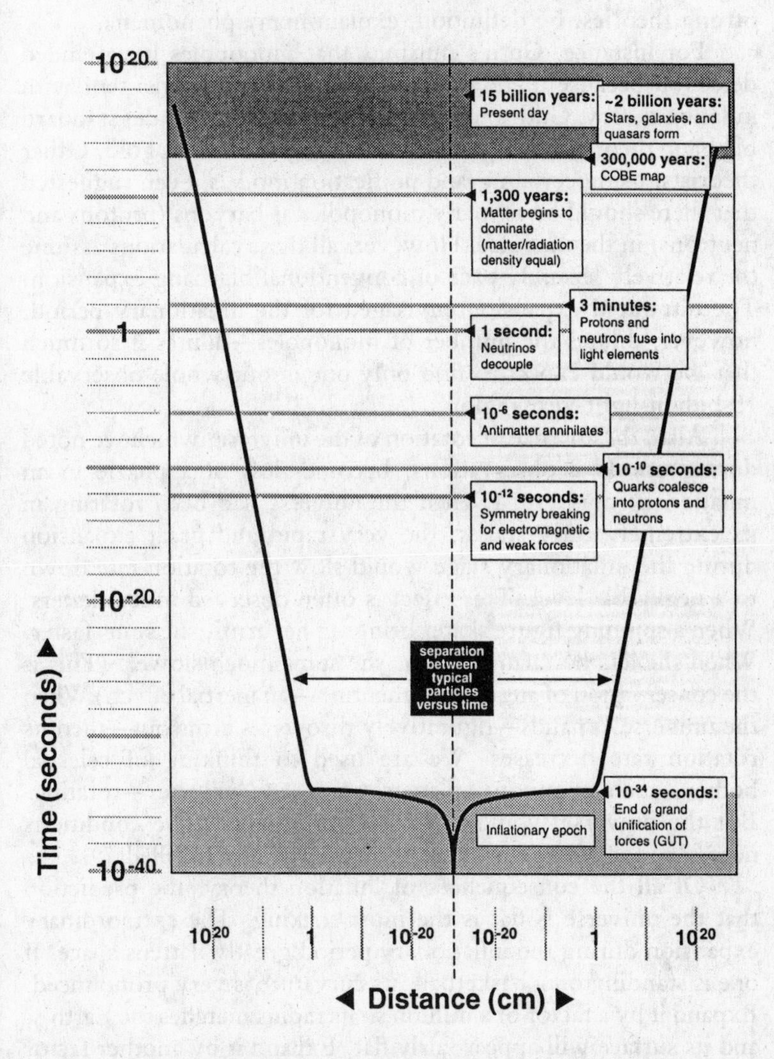

trary assumption of initial conditions, nor contact between disparate regions of the universe. It merely proposes that the initial, inevitable homogeneity of matter became the universal condition through brief, explosive growth. If this were all that inflation theory implied, then, elegant though the solution may be, it might be judged weak. But it touches on many cosmological problems. Strong theories, by definition, explain many phenomena.

For instance, Guth's musings that monopoles have eluded detection because there are very few of them is consistent with inflation theory. Guth and Tye had calculated that, under standard big bang theory, monopoles would be as common as gold. Other theorists, using certain grand unification models, even suggested that there should be as many monopoles as baryons (protons and neutrons) in the universe. However, all these calculations assume the relatively leisurely pace of conventional big bang expansion. The ultrarapid expansion envisaged for the inflationary period, however, dilutes the number of monopoles—dilutes it so much that we would expect to find only one in our whole observable 15-billion-light-year region.

Also, the absence of rotation of the universe, which we noted during our U-2 observations, becomes less of a puzzle in an inflationary universe. Even if the universe had been rotating in its extremely early stages, the very rapid and great expansion during the inflationary stage would slow the rotation rate down to a negligible level. This effect is often observed in ice skaters. When a spinning figure skater brings in her arms, she spins faster. When she lets her arms fly out, she spins much slower. (This is the conservation of angular momentum—an inertial effect.) When the universe expands—figuratively throws its arms out—then its rotation rate decreases. We are used to thinking of celestial bodies—such as moons, planets, stars, and galaxies—rotating. But the universe would not rotate noticeably. The conditions necessary for Mach's principle will be partially fulfilled.

Of all the consequences of inflation theory, the prediction that the universe is flat is the most striking. The extraordinary expansion during the inflationary period greatly flattens space. If one is standing on a basketball, its curvature is very pronounced. Expand it by a factor of a million so its radius matches the Earth's, and its surface will appear fairly flat. Expand it by another factor of a million, and its surface will look essentially flat. Inflation adds much more than a million-times-a-million expansion. Flat space corresponds to a prediction that Omega—the ratio of the critical density of the universe to its actual density—must equal 1.

As the balloon is inflated by orders of magnitude, the surface curvature in any given region—say, the region around the bug—quickly becomes undetectable (though present). CHRISTOPHER SLYE

As we saw in the previous chapter, the value of Omega can be, and has been, measured, thus testing the validity of inflation theory. At present, a reasonable range estimate for Omega is 0.3 to 2, with many astronomical measurements indicating something close to 1. This may look approximate in the extreme, but given the constraints on cosmological measurements, it is a very encouraging match. Had our estimates been off by an order of magnitude or two, then inflation theory would have done poorly in this test, and perhaps be declared as having failed. We can say, therefore, that the Omega-equals-1 prediction has a fair chance of being vindicated. Linked to this prediction, of course, is the further prediction that much of the matter in the universe must be the enigmatic dark matter. The visible matter in galaxies accounts for a fraction of 1 percent of the critical mass; so, if the critical mass is to be 1, most matter must be dark. For inflation to remain a viable model, astrophysicists must find the dark matter or very strong evidence of its existence.

The inflation concept looks extremely successful as a solution to a range of cosmological problems. But it would not have captured so much attention had there been no mechanism to make it work. The mechanism was found—the breakthrough achieved—by uniting ideas from diverse areas.

In the last week of June 1979 I attended a symposium on "The Universe at Large Redshifts" in Copenhagen at the Niels Bohr Institute. As Dennis Sciama and I walked across town from our hotel to the institute one morning, we talked about the cosmic microwave background and Mach's principle—how they were related and what might have linked them. We were having such an enjoyable time and were feeling so close to some link that we agreed to lunch together and continue the discussions. It was a beautiful day, so, joined by Steven Weinberg, we got our smørrebrød, sardines, and Carlsberg* beer and went out to sit in the grass under the sky and trees. Soon Weinberg asked us, "What do you know about phase transitions?" I said, "I only know what I learned in Philip Morse's thermodynamics course and from junior-year physics lab. I remember the phase transition experiment vividly—as the liquid passed the critical point, it got all cloudy and disoriented and then, after a brief moment of calm, froze in a flash."

*The Carlsberg Foundation was a primary symposium sponsor and has had a long interest in science. The brewery had run a direct line to Niels Bohr's house so that he always had their inspiration on tap.

Stephen Hawking (*left*) and George Smoot at a conference in Berkeley, December, 1992. LAWRENCE BERKELEY LABORATORY

"Well," Weinberg said, "I am working on phase transitions in the early universe and need to find out if the GUT phase transition is first order [thus discontinuous, like water to ice] or second order [smoothly continuous]. Phase transitions are important."

Our discussion turned back to the cosmic background radiation. Sciama remarked how appropriate it was that we were at Bohr's house, because Bohr was one of the founders of quantum mechanics. And quantum mechanics, Sciama asserted, would provide not only the explanation of the origin and isotropy of the cosmic background radiation but also a way of escaping the initial singularity, the black hole at the beginning of the universe.

Singularity theorems from Roger Penrose and Stephen Hawking inform us that if one extrapolates back far enough in time, then our space-time must come to a geometrical point—a catastrophe for us and the theory. However, as we look back, the matter and space-time of the universe get cramped into smaller and smaller volume, making quantum mechanical effects more and more important. The quantum mechanical (Heisenberg) uncertainty principle tells us that both matter-energy and space-time must fluctuate. The matter-energy fluctuations, like struggling in quicksand, just make things worse, increasing the effect of gravity. However, Sciama pointed out, the curvature fluctuations of space-time do the opposite, thereby weakening gravity's effect. Weaker gravity could violate the conditions needed for the singularity theorem to be true and allow escape from the initial singularity. Hawking had calculated this is what happens near a very small black hole that allows the occasional emission of particles. This emission from black holes is called Hawking radiation. Sciama reckoned that quantum effects, through the Hawking emission of energy and particles during the quantum phase of cosmology, could be the origin of the cosmic background radiation and could avoid the singularity—neatly killing two birds with one stone. This lunchtime conversation touched on all the elements of inflation, an idea whose time had clearly come.

Nearby, Katsuhiko Sato, on leave from Japan to work in Copenhagen, would soon realize, as would Guth a continent away, that particle physics' grand unification might lead to an inflating universe. He also realized something about inflation that Guth failed, at first, to see: that the inflation that smoothed out the early universe would also produce small fluctuations, which might lead to galaxy formation. However, Sato's paper was slow in being published; though submitted to a journal in early 1980, it did not appear until late 1981. By one of those curious coincidences, the stage was set for this realization to burst out on the cosmological community.

In the summer of 1982 Stephen Hawking organized a conference on the early universe at Nuffield College, Oxford, England, bringing together Alexei Starobinsky, Alan Guth, So Young Pi, James Bardeen, Paul Steinhardt, Michael Turner, and others. They quickly focused on the idea that quantum fluctuations could produce perturbations within the uniformity of the inflation bubble. A quantum mechanical fluctuation—a tiny wrinkle in space-time—produced early during inflation would be stretched to a tremendous length by the same expansion that was making space so uniform and big. A wrinkle generated slightly later in the expansion would not be stretched quite as much, and so on. If the expansion rate of the universe was constant, then the quantum mechanical fluctuations would all have the same characteristic wrinkle height but would have a great diversity in length, depending on when they originated. Inflation does not erase the quantum mechanical fluctuations, but establishes them as macroscopic, cosmological-size ripples spread through space-time. Some will be of just the right size to spawn structures we see today. These wrinkles result in the gravitational accretion of visible and dark matter, forming structures, from galaxies to superclusters and beyond.

How must the primordial wrinkles have been distributed in order to explain current cosmic structures? Back in the early 1970's, Edward R. Harrison, of the University of Massachusetts, and Yakov Zel'dovich, in Moscow, had independently considered this question. They concluded that the distribution of celestial objects we see—galaxies, clusters, superclusters, and other gargantuan structures (such as the Great Attractor)—would be produced only if the distribution of fluctuations does not depend upon their physical length. If there were many more small fluctuations than large ones, the universe would swarm with black holes, with few structures as large as superclusters. Conversely, a preponderance of large fluctuations would lead to a universe dominated by megastructures. In the latter case these various celestial behemoths' gravitational attraction would whip our Milky Way (and other galaxies) through space at velocities that would make our present peculiar velocity of six hundred kilometers a second look like an afternoon stroll. Think how the astronomers would have reacted to that, given their prejudice against any peculiar motion of galaxies!

By now it should be apparent that inflation is the most influential concept in modern cosmology. Although it doesn't solve all questions about the origin of the universe, it provides persuasive solutions to many. The question now is, How did this big bang within the big bang occur? How could a brief, explosive expansion, which was so different from the normal expansion of the big bang, occur in the early life of the universe? It is here that Guth has pointed to a harmonious marriage between astrophysics and particle physics—a match so beautiful it has to be right.

The early universe was extremely hot and extremely dense and thus had a high energy density. If we assume it was expanding at an enormous pace, its expansion rate will decline because of gravitational attraction. Unless, that is, the newly created space comes with its own energy density, enough to outweigh the matter-energy density, thus keeping the expansion process going at an accelerating rate. How does this happen?

Guth appealed to the grand unified theories for inspiration. At the end of the GUTs era, at 10^{-34} seconds, the symmetry breaks, yielding differentiated strong and electroweak forces. Before this symmetry was broken, the forces and matter were unified, possibly endowing the primordial vacuum with vast energy density. As the universe expanded at this early instant, more space was created, also endowed with energy density. Expansion could therefore continue at a constant rate: The scale size of the universe might double about every 10^{-38} seconds. At that rate, in 10^{-35} seconds the scale of the universe doubles a thousand times. While sounding innocently mild, doubling a thousand times shows the power of inflation, or compound interest, and makes the universe 10^{301} times bigger than it was in less than an eyeblink.

If this is how inflation is initiated, how does it stop? How is the transition made between a period of accelerating expansion and one of continuing but decelerating expansion? In 1982 Paul Steinhardt and Andreas Albrecht, at the University of Pennsylvania, and Andre Linde, in Russia, came up with a way. This transition is important since the tremendous expansion of inflation has made space extremely cold and empty. When the energy density drains from the vacuum, it goes into particles and energy. At this point, gravity begins to exert its attractive effect, gradually slowing down the expansion. The accelerating inflationary phase thus comes to a halt. We can now see why our present universe is expanding as it is: We are in the coasting aftermath of the accelerating phase of expansion.

If you contemplate the inflationary scenario carefully, you come to another remarkable insight. The condition for inflation to occur is that the total energy—that is, the energy of space itself minus the gravitational attraction of the other parts of space—be essentially zero. Therefore, we can still conserve energy and make everything in the universe starting from practically nothing. All inflation needs is a small region in the right configuration and it will run away to produce a bubble in space vastly bigger than the whole of what we can observe today. You can see now that inflation gives new meaning to the philosophical question of "nothingness and being."

But how did conditions become right in that small region? This is still an unsolved mystery. There are theories, of course. Some people—for example, Andre Linde—say that the universe is chaotic and the natural scale is the fully unified scale, so that there may be very many inflation bubbles and chaotic space. Others, such as Stephen Hawking, theorize a definite, well-defined beginning. The question of "the beginning" is as inescapable for cosmologists as it is for theologians.

Meanwhile, let us relish the power of so elegant a concept as inflation, which promises to solve so many cosmological puzzles and gives us so clear an insight into the tenuousness of our existence. When we learn of the consequences of Omega being anything other than precisely 1, we see how very easily our universe might not have come into existence: The most minute deviation either side of an Omega of 1 consigns our potential universe to oblivion. And when we see that a noticeable tilt in the quantum fluctuation spectrum of primordial ripples might have produced, instead, a vast swarm of black holes, or a cosmos of lumbering giants, then we again realize how easily things might have been very different. Unless, as inflation theory holds, initial conditions and early processes *had* to be as they were.

These observations, particularly the requirement that Omega be exactly equal to 1, strike some theorists as so unbelievable that they've turned in desperation to the anthropic principle, a term coined in 1974 by Cambridge University cosmologist Brandon Carter. The idea has several guises. Each relates to the notion of us as observers, and to the conditions necessary for a universe in which we can exist. There is a long list of physical laws and conditions that, varied slightly, would have resulted in a very different universe, or no universe at all. The Omega-equals-1 requirement is among them. Perhaps numerous other universes formed in which Omega does not equal 1, but they're lifeless. Hence our Omega equals 1 only because if it didn't, we wouldn't be here to notice the fact. To Guth and other skeptics, myself included, the anthropic principle is unsatisfying. "The anthropic principle is something that people do if they can't think of anything better to do," Guth once joked. The dangers of this view are illustrated by the Parable of the Fish Philosophers. These denizens of the dark and briny deep periodically met to impress themselves and each other with their insights and bioluminescence. One particularly brilliant fish philosopher had concluded that space—the water around them—was homogeneous and symmetric and also warm, full of nutrients, and just the right pressure for life. Had things been different they surely could not have existed. Their bodies would either implode or explode if the pressure were different. Each chimed in with another example until the list was quite long. One among them chimed in with the "fishtropic principle." The fish philosophers floated around congratulating themselves that through such cleverness their efforts cast a light through the murky deep. Swelling with pride, they did not notice a fishing net had closed around them, inexorably drawing them upward. "Hey, fellows," the marine biologist operating the net called to the other scientists, "we caught a big load of the poor devils this time. They must have been having a school or workshop." His colleagues didn't hear. They were awestruck by the brilliant lights moving so rapidly in the sky. They didn't even notice their ship lifting . . . Saying that things must be as they are simply because we exist can lead to a fatally provincial and narcissistic view of the universe.

Many things that cosmologists thought in 1974 were miraculously fine-tuned to allow life and human existence are neatly and powerfully explained by inflation. In their book *The Early Universe,* Rocky Kolb and Mike Turner state: "It is unclear to one of the authors how a concept as lame as the 'anthropic idea' was ever elevated to the status of a principle." I agree. I think that a fuller set of observations will lead to models and theories that will gracefully and easily explain why things are the way they are. Whatever these future discoveries and concepts are, I am confident that, like inflation, they will astound and delight us with their elegant simplification and unification of nature.

Despite its scope and power, inflation theory cannot show us the distant echoes of creation that it so cogently supports. The theory explains why the cosmic background radiation is so extraordinarily smooth, as we and others have observed. The task of finding signs of the minute perturbations in the background—which must exist if our conception of the universe is correct—falls to the experimentalist.

Chapter 10

The Promise of Space

In 1974, the era of cosmological satellites began. NASA issued dual announcements of opportunity, No. 6 and No. 7, which invited researchers to propose astronomical missions for small and medium-class *Explorer* satellites. This was the chance I had been looking for. In truth, it was something I had dreamed of since I was a kid, reading Arthur C. Clarke's books on engineering and science and his science fiction. He invented concepts such as the geostationary satellite, which he said could be used for communications and entertainment, and proposed an orderly development of the space program. Clarke was a visionary with a vivid imagination, and his vision fired me with the wonder of space as the new frontier for exploration. In 1957, when the Soviets launched *Sputnik,* I knew the future that Clarke had prophesied could be real. Soon I was listening to the *beep-beep* and dreaming of the day when I could be working in space. Yes, it was a romantic fantasy, but even as a kid I knew space was where ground-breaking science one day would be done. Now, two decades later, NASA's announcement had brought the prospect to reality. I could barely believe my good fortune.

At the time I had been in cosmology for just four years, convinced that that was where the fundamental questions would be asked and could be answered. I'd had my first experience lofting exquisitely sensitive scientific instruments into the stratosphere on gigantic balloons, all too often to have them finish up as scrap metal or lost forever. And I was just embarking on the thrill of turning a spy plane—the legendary U-2—into a high-flying observatory, to search for clues to the dynamics of the universe by exploiting the cosmic background radiation. But satellites—yes, that's what I really wanted. I would be immersed in the technical challenges of designing an instrument that could work in space; I could realistically expect data that would be qualitatively superior to data from ground-based, balloon-based, or plane-based observatories, and therefore win new insights into cosmological issues; and, through it all, the romance would be sweet.

Or so I thought. I have to admit that, when I was formulating a response to NASA's 1974 announcements—enthusiastically encouraged by my group leader, Luie Alvarez—I had no idea of the difficulties and heartbreaks that lay ahead. Perhaps it was a good thing I didn't.

I realized that for a while I would have to juggle the development of ideas for satellite-borne projects with the immediate task of getting the U-2 venture under way. That was just fine, because I viewed a satellite experiment as the natural successor to the work we would do in the stratosphere on the spy plane. Each project could learn from the mistakes and successes of the other.

With my Berkeley colleagues, I immediately began drafting a proposal—in effect, two in one—for cosmological experiments that could be launched on *Explorer* satellites. Both concerned aspects of the cosmic background radiation. The first experiment would make a map of the microwave sky to search for signals in the cosmic background radiation, evidence of the elusive wrinkles. The second experiment, requiring a large instrument, would measure the spectrum of the background radiation, to determine whether it was truly the relic radiation from the big bang with the expected thermal, or blackbody, curve. I guessed that the smaller experiment was most likely to succeed in the competition, and so we focused on it. The spectrum-measuring instrument grew so large that we decided to submit it as a mission by itself—as a satellite to be lofted on the space shuttle, to be tossed overboard in space.

For our instruments, there was a tremendous advantage to being in space. First, there would be no atmospheric noise to interfere with the measurements and no weather to disrupt the experiment. We could look at any and all wavelengths needed without the obscuring atmosphere. We could make calibrations easily, since we would have no worry about condensing water or freezing atmosphere. The satellite's orbit and rotation would automatically scan the sky. Space provides an easier-to-control and thus more even-temperatured environment than any on Earth. Being in orbit meant there would be sufficient observing time to gain the needed sensitivity and calibrations. Things would be a lot better in space.

I opened new connections with Ball Brothers Research Corporation, with whom we had been working on the High-Energy Astrophysical Observatories satellite, an extension of our antimatter and cosmic ray balloon program at Berkeley. Then I made links with Hughes Aircraft Spacesystems, and with AeroJet ElectroSystems, particularly a group led by Herb Pascalar. I needed the expertise of the three companies to learn how to convert from aircraft- and balloon-borne instrumentation to a spaceborne version. Doing science this way seemed to me more like being a technological entrepreneur than a scientist—nevertheless, there was no question it was intellectually demanding. For instance, during the proposal writing Tim Tyler of Ball (who afterward left to prospect for uranium) and I developed a new methodology for accounting for instrument performance and especially the things that could go wrong—systematic errors. A large part of what I was to do for the next twenty years was to continue and upgrade this effort, as well as keep up on the science.

Meanwhile, Samuel Gulkis, Mike Janssen, and their colleagues at the Jet Propulsion Laboratory (JPL) in Pasadena were working on their own proposal. They, too, thought a small *Explorer* experiment would stand the best chance for approval and funding. The experiment they proposed was also a cosmic background radiation mapper. Their design was not for a differential microwave radiometer, but for a total power radiometer on a high-orbit, rapidly spinning spacecraft. This was a good concept but ahead of the available technology at the time.

In New York City, unbeknownst to both groups, Pat Thaddeus mentioned the NASA announcements to John Mather and suggested: "Why don't you see if you could come up with something?" John and Pat teamed up with Rainer Weiss and Dirk Muehlner of MIT, David Wilkinson of Princeton, and Mike Hauser and Bob Silverberg of the Goddard Space Flight Center in Maryland. They proposed a cryogenically cooled cosmological satellite that, at the time, seemed to be "a completely crazy idea," John acknowledges. "What was crazy was we thought we knew what we were doing," he explains. The proposal included a cosmic background mapping experiment (like ours), advocated by Dave Wilkinson and Ray Weiss; a cosmic background radiation spectrum measuring experiment, which was an extension of John Mather's doctoral thesis instrument; and a small telescope to measure the cosmic infrared background, which is the glow from the earliest luminous objects (presumably stars and galaxies).

NASA received more than 120 proposals from across the nation. Those of us who were cosmologically inclined were a small subset: Only three groups—mine, Sam Gulkis's, and John Mather's—advocated launching satellites to monitor the cosmic background radiation. My group's proposal for searching for wrinkles in the cosmic background radiation called for launching a small instrument aboard a small, bargain-basement rocket called the Scout. So did the Gulkis team's proposal. John's proposal demanded a deluxe satellite.

My shuttle-borne spectrum proposal was turned down as too large and too risky. John's team, with a smaller spectrum and diffuse infrared background experiment, was told to join a group planning the Infrared Astronomy Satellite (IRAS). It appeared as if it would be a shoot-out between JPL and Berkeley over whose proposal would be accepted. Or our two groups might be put together in a shotgun wedding—a favored procedure of NASA headquarters at the time. Before the finale, lack of real estate on IRAS caused John's group to be unceremoniously ejected from the IRAS group, and so he was back looking for a route to space. NASA headquarters decided the three groups—Gulkis's, Mather's, and mine—should work together. In 1976, NASA appointed six of us (Sam Gulkis, Mike Hauser, John Mather, Rainer Weiss, Dave Wilkinson, and me) to guide a study team to evaluate and develop our ideas. The study group was augmented by other members of our proposal teams and by engineers and managers. Mike Hauser convinced John Mather to move to the Goddard Space Flight Center in Greenbelt, Maryland, from New York City, and their combined efforts convinced the center to take on responsibility for the study and mission.

We proposed a mission named the Cosmic Background Explorer, or COBE, bearing three instruments. We proposed to use differential microwave radiometers (DMRs)—like the one used on the U-2, but more sensitive—to map the universe as it appeared roughly three hundred thousand years after the big bang and to look for the primordial cosmic seeds. I would be the DMR principal investigator. John Mather would lead the far infrared absolute spectrophotometer (FIRAS) project, which would measure the spectral curve of the cosmic background radiation. The spectral curve shows the intensity at each wavelength of the radiation, and its overall shape would tell us whether the radiation had been produced by the big bang event, or by something else. And Mike Hauser would head the diffuse infrared background experiment (DIRBE), which would look for the cosmic infrared background, the glow from the earliest luminous objects such as primordial stars and galaxies. These might date from the first tens or hundreds of millions of years after the big bang.

The scope of COBE—if approved—would be, in effect, to assemble a series of "baby pictures" of the newborn cosmos, taken at different times to show different steps in its evolution. It would be a major adventure in cosmology, and would surely springboard the science to new levels of technical achievement. We had a clear idea of the science we wanted to do, and a notion of the technical challenge of implementing it. NASA officials, for their part, were well used to scientists whose ideas ran ahead of the very stringent demands of getting instruments to operate smoothly with no hands-on intervention in the cold void of space. Before they would commit themselves, we would have to convince them that we knew what we wanted to do and knew how to make the instruments to do it. It was a slow process.

Six years would pass.

At Berkeley, we continued with our research. We conducted antimatter experiments, we observed the cosmic background radiation from atop White Mountain, we launched the U-2s from northern California and Peru—and we worked steadily on the COBE concept. I began commuting back and forth between Berkeley and NASA's Goddard Space Flight Center, and started to form a core team for the DMR instrument. It was the beginning of a long journey. When at Goddard, I spent part of my time in Building 2 working with the other COBE scientists on the mission concepts and part in Building 19, where I looked for a few good engineers. First on the spot were Roger Ratliff and John Maruschak—a duo who formed the engineering nucleus of the DMR team at the Goddard. I educated them about the scientific rationale for the experiments and the requirements for the DMR instrument. They educated me about the engineering requirements for space missions.

From the U-2 and ground-based work I knew that testing and calibrating the new DMRs was going to be a major problem. We couldn't afford to take the risk that, once in orbit, the DMRs wouldn't work as they should. NASA officials, quite properly, wouldn't allow us to take such a risk. They insisted that we prove that the instrument would work under spaceflight conditions.

NASA must have had a deal with fate. As we were wrestling with how best to test the DMR, we were startled by an announcement that sent us scurrying to get the instrument aloft as fast as we could—and by whatever means we could.

In 1980, news whipped around the cosmological community that two scientific teams—one Italian, the other American—had made a major discovery. According to gossip, they had detected the first evidence of cosmic anisotropy that was intrinsic to the cosmic background radiation, rather than related to the motion of Earth and our galaxy. The Italian team was led by Francesco Melchorri, then at the University of Florence, and the American team by David Wilkinson, a pioneering cosmic background radiation researcher at Princeton. Both teams had launched balloon-borne instruments, and both had detected possible evidence of a cosmic quadrupole. It was stunning news.

A dipole has two poles, the warmest and the coolest parts

of the sky, which results from our motion relative to the cosmic background radiation. This is the dipole detected by the U-2. A quadrupole is a pattern of temperature on the sky with four poles—two warm and two cool, and in a symmetric pattern. A quadrupole could be caused by many things, all likely to be important. For example, it could be evidence of cosmic rotation or of the universe expanding more rapidly in one direction than another. One would also get a quadrupole temperature distortion if a very long wavelength gravity wave were passing through this region of space. Or—the most intriguing possibility of all—a quadrupole and many other higher-number pole fluctuations could be due to the effects of the long-sought wrinkles. That is, it could be cosmic background radiation anisotropies marking primordial density fluctuations—the seeds of future galaxy clusters. Our U-2 data had failed to detect such a quadrupole, but the telltale signal may have been just slightly fainter than our measurements could indicate.

Had the Wilkinson and Melchorri teams found the Holy Grail of cosmology before we had even started looking?

We reacted cautiously—even equivocally—to the reports. On the one hand, we noted that both teams had collected data at wavelengths where the emission from our galaxy is a significant contaminant. The quadrupole therefore might be an illusion caused by microwave radiation emitted by the Milky Way galaxy. On the other hand, we said in an article for *Physical Review Letters:* "Current theories suggest a natural interpretation for this type of anisotropy as arising from density fluctuations"—the long-sought-after cosmic seeds that would grow to galaxies, clusters, superclusters, and clusters of clusters.

I would have been delighted if Wilkinson or Melchorri had found the seeds. After all, by the early 1980's the failure to detect ripples had begun to worry cosmologists seriously. Even as detectors improved in sensitivity, they still failed to detect temperature variations at a level of one part in a hundred. Then our U-2–borne experiment got down to one part in a thousand and found the dipole anisotropy due to our motion, but nothing else. Now sensitivities were pushing down toward one part in ten thousand, and still nothing. The formation of galaxies through gravitational attraction required the existence of primordial seeds and thus fluctuations in the cosmic background. Cosmological theory demanded that fluctuations exist, and inflation provided a way for them to be formed. Their eventual discovery was going to be a major event in the history of cosmology and, though I really wanted to be the one to discover them, I was ready to applaud their discovery by others.

But before applauding we had to be sure that the newly discovered quadrupole was real. I had planned to continue the U-2 experiment with a new and better receiver. Back at the lab—LBL and the Space Sciences Laboratory—we had built a new liquid-nitrogen-cooled receiver to fit in our U-2 instrument. The technology of quiet radio receivers had advanced: Cooling receivers produced lower instrument noise. It was natural to use them for improved measurements. I had submitted a proposal to NASA to fly these cooled receivers on the U-2 before the announcement of the quadrupole. Now I pushed to get approval so that we could look and check the quadrupole. There were more delays. I was frustrated.

Simultaneously, worried theorists pushed their ideas to the limit. They had been revising their theories to account for the inability to detect the ripples, but one can rewrite theories only so much before reaching fundamental limits. If gravitational attraction formed structure from primordial baryonic density fluctuations, then there must be fluctuations by one part in ten thousand. If cold dark matter existed at the critical density, it was possible to push down to one part in about one hundred thousand. One could not go much lower and still have gravity make the structures we see, from galaxies to superclusters. If we did not find fluctuations at a level of a few parts per million, we would have to throw up our hands and walk away, because some new force would be needed and we would have no notion of what was happening in the early universe—it would be a new game and we would have no idea of the rules.

The new urgency we faced was rewarded with further frustration: Planning for COBE was delayed, primarily because the IRAS satellite was delayed, holding up the waiting line in which COBE stood. Worse, there was no immediate prospect of getting our new cooled differential microwave radiometer on a U-2 flight anytime soon. How were we going to test the instrument to NASA's satisfaction?

There was only one choice—balloons.

Given my earlier experience with balloons—and my faded romance with them—I would have taken any viable alternative.

Grumbling at my fate, I nevertheless focused keenly on what had to be done. We would look for the putative quadrupole at a different wavelength, three millimeters, where galactic contamination is substantially reduced. If we saw the reported quadrupole through this clearer galactic window, we would have a much stronger reason for believing that cosmology was on the brink of a crucial discovery—even if it wasn't the ripples, it would be something important. We planned to conduct four balloon flights—two from the Northern Hemisphere and two from the Southern Hemisphere. If the two flights in each hemisphere were separated by a few months, we would have nearly full sky coverage.

I discussed the project with John Gibson, our electronics engineer, and Hal Dougherty, our mechanical wizard, both of whom had previously worked on the antimatter and U-2 experiments. They got to work on the basic gondola and electronics. Next I had to find a good graduate student. My U-2 colleague Marc Gorenstein had already finished his Ph.D. and left to study quasars at MIT. So I searched around and quickly located an outstanding student—Gerald (Jerry) Epstein. He had earned a joint degree in physics and electrical engineering at MIT, then came to Berkeley to get his doctorate in physics. He was initially reluctant to get involved with the balloon experiment—and I can't blame him. His goal was to work as a science policy analyst for the U.S. government (as he eventually did, working at the Congressional Office of Technology Assessment). But I persuaded him that he could render better judgments on national science policy after doing frontline research of his own. Going out into the field and dealing with weather, strange people, cranky equipment, data analysis, and so on would be a broadening experience. Jerry agreed to join the experiment, and I think he profited as a result.

About that time, I got a phone call from Philip Lubin, my former graduate student who had been involved with the U-2 flights in Peru. Phil wanted to develop a three-millimeter-wavelength receiver that was cooled with liquid helium at four degrees above absolute zero: 4°K. This receiver would make the instrument two or three times more sensitive than the liquid-nitrogen-cooled detector I had developed at Berkeley. Phil and I agreed to collaborate on a joint balloon project.

We worked hard and fast. I worked out the specifications and connections for a liquid-helium dewar (a special container for liquefied gases), then ordered one from a firm called Infrared Labs. Hal designed and constructed for the dewar a snout and a window through which one of our special horn antennas looked at the sky. The antenna had to look through a window because its outer part was cooled by liquid nitrogen to about 80°K while the inner part was near 4°K. At those temperatures, air would freeze onto the antenna. Jerry worked with Hal to design the framework, and with John to develop the new electronics for the detector. Phil worked on the three-millimeter receiver. We planned the experiment and made sure that we understood all the possible sources of radio noise, such as the atmosphere, the Moon, stray Earth signals—a whole long list I had developed for the U-2 and COBE DMRs.

A key problem was how to "switch" the signal. A differential microwave radiometer measures the differences between temperatures in two parts of the sky. Our previous DMRs—operating at room temperature or with liquid-nitrogen cooling—switched from one part of the sky to another using electromagnetic switches that were sure and reliable. But little was known about switching technology for detectors operating at the much colder temperatures of liquid helium. We thought of many different schemes for a switch, but none was very good. And we needed something quickly.

This led to our development of the "salami slicer." Our cosmic background radiation spectrum measurements used moving aluminum mirrors to reflect the beam from different angles on the sky. In that same way, we had ensured the DMRs aboard the U-2 were giving a zero output, as expected, for the difference of identical signals. Then I had the idea of using a similar method to make the kind of high-speed switching between parts of the sky necessary for our new instrument: We would tip our detector to view the sky at 45 degrees from zenith. If in front of the detector we could rotate a vertical aluminum mirror with big slices cut out of it so that it resembled a propeller, the beam would switch between two parts of the sky 90 degrees apart. It would see the "true" sky through a gap in the mirror, then the reflected sky, then the true sky again, then the reflected sky, and so on, as the mirror's propellerlike blades rotated. This process is called chopping. By quickly spinning the mirror, we could chop very rapidly—rapidly enough to achieve the high-sensitivity measurements necessary. Also, the 90-degree separation allowed us to achieve maximum sensitivity to a quadrupole anisotropy, where the warm and cool poles are 90 degrees apart.

A motor slowly rotated the gondola—turning the balloon one way and the gondola the other (in a way we hoped would not twist up the parachute). The beam would scan out a circle on the sky, and the rotation of the Earth during the night would cause the circle to move across the sky. Because all measurements were made at 45 degrees from vertical, the signal from the atmosphere would be identical and would cancel out in the difference measurement. In short, we would be in the map-making business. All we needed was a big, rapidly spinning mirror.

We all threw ourselves into building the mirror chopper system. Hal cut a two-foot-diameter mirror out of a sheet of highly polished aluminum. Then he mounted the glistening "propeller" on a motor that John Gibson had wired up. Jerry and John figured out how to synchronize the detection of the signal and the rotation of the mirror. We programmed the equipment to raise periodically a small target emitting a signal of known intensity; if the detector measured the same emission level each time, then we'd know it was working fine. Soon the whole machine was set up in the lab. We turned on the switch; the mirror started rotating. It picked up speed until it reached its design 700 rpm, where it purred and whistled as it sliced through the air. We had a chopper, and it did look and sound like a giant salami slicer. A couple of times we got too close to it and narrowly

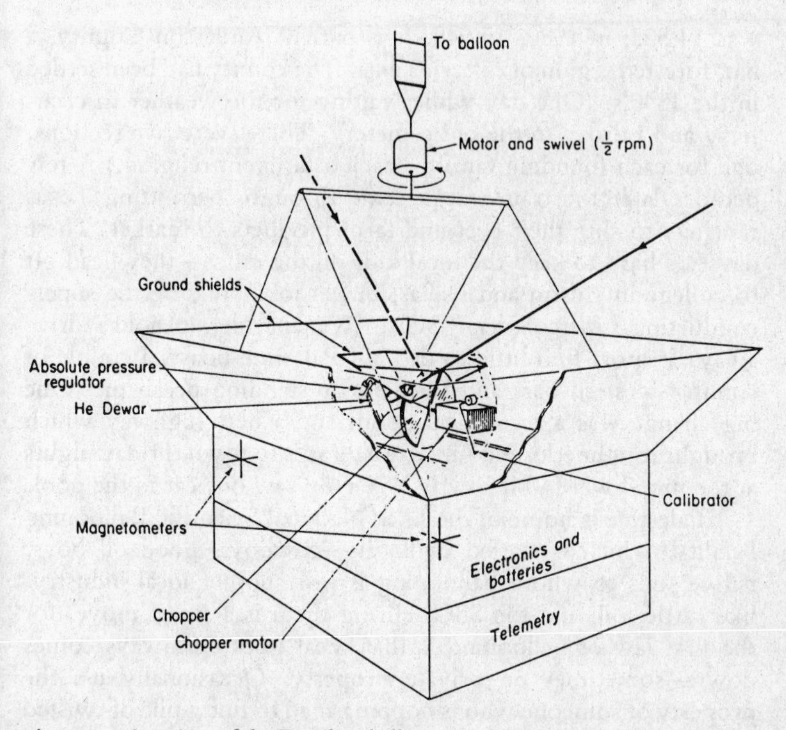

A perspective view of the Brazilian balloon gondola. Note the chopper. LAWRENCE BERKELEY LABORATORY

missed being gouged. Around it we placed a protective barrier, very much like a chain guard on a bicycle.

We completed the basic DMR very quickly. Phil finished making the three-millimeter receiver, and he and Jerry mounted it and its Jerry-built amplifier into the dewar. When everything was ready, we filled it with liquid helium and took the whole package, salami slicer and all, to the roof of Building 50 at Lawrence Berkeley Laboratory for tests. It worked perfectly. Now all we needed was an opportunity to fly it on a balloon.

Palestine (the last syllable is pronounced "steen") hadn't much changed since my last visit there, when we were launching antimatter experiments in the early 1970's. Its population in 1970 was 14,525; in 1980, 15,948. It is part of Anderson County, a flat, forested region of eastern Texas. The county had been settled in the 1840's. (One day while waiting for the weather to clear, Jerry and I drove to the old cemetery. There were five sections, one for each founding family—each a different religion.) A few decades later the train tracks came through, permitting Texas ranchers to ship their beef and farm products to market. These days it's hard to keep the local kids on the ranch—they head off to college in Austin and Dallas, or get jobs building the superconducting supercollider (SSC) in Waxahachie, an hour's drive away. Visitors find little to do, but Palestine boasts a couple of country-western bars and Lucille's outstanding pecan pie. One big change was a second oil boom and a new highway which brought roughnecks and more restaurants to town. Friday nights at the motel were a bit rowdy. We only saw one car in the pool.

Palestine is home of the U.S. National Scientific Ballooning Facility, which is staffed by locals—cowboys, good ol' boys, native sons to whom ballooning is just another local industry, like cattle, oil, and the SSC. Hiring them is a smart move, for the first law of ballooning is that what goes up always comes down—sometimes on private property. Occasionally it's the property of someone who is hopping mad to find a pile of twisted metal and flapping polyethylene on prime farmland. Now and then a distressed landowner will call the sheriff. In such situations a homegrown recovery crew is indispensable.

The balloon base is busiest in early summer (May–June) and autumn (September). Scientists prefer to launch their payloads at those times because the high-altitude jet streams reverse directions. During those times the winds, though tending to blow in random directions, are usually low in speed, hence a balloon is likely to stay in the same area and needn't be chased many hundreds of miles, sometimes clear to the Gulf of Mexico, as occurs when the prevailing winds are steadier.

In July 1981, Dave Wilkinson was flying a balloon-borne anisotropy experiment when we headed for Palestine, and he graciously consented to let our experiment ride along for an engineering checkout. Four months later, after we'd made adjustments as a result of that first outing, we hitched a second ride, this time on an MIT balloon carrying the Ph.D. thesis experiment of Mark Halpern, a graduate student working with Ray Weiss. After an eight-hour flight at an altitude of thirty kilometers, the gondola was released, and subsequently landed in Mobile, Alabama, around dawn. Apart from some local consternation at whimsical signs taped on the gondola—MIT-BERSERKELEY LOONEY LANDER and PROPERTY OF THE U.S. AIR FORCE—recovery went relatively smoothly. We were amassing data that would eventually check the quadrupole claims.

The next flight was uneventful. This time we used our own newly built gondola—no more "hitchhiking." Our payload ascended into the sky at 7:45 P.M. local time, on April 26, 1982, and rose to more than twenty-eight kilometers. It drifted east, then turned south eleven hours into the flight, and descended near Baton Rouge. It landed in an irrigation ditch between two fields. As the crew pulled the gondola out of the ditch, Jerry popped its data tape—which resembles an ordinary audiotape—into the back of his carrying bag. He also stuck in the bag the small motor that rotated the gondola under the balloon. All in all, a surprisingly smooth flight.

That night, Jerry suddenly awoke in his motel room. Lying there in the darkness, half-awake, he had a vision of the interior of his bag, within which lay the tape and the motor. Suddenly he remembered that motors contain magnets, and magnets destroy electronic data. He leaped out of bed, switched on the light, fumbled through the bag, and snatched out the tape. There was no way of knowing whether the flight data had survived until he could get it back to Berkeley and put it in the computer. He imagined our reaction if he had to confess how months of work had been ruined by a single careless maneuver.

"I was one nervous person," he recalls. "When we got back to Berkeley and analyzed the tape, I didn't breathe until we'd reviewed a few minutes of data. It turned out to be fine."

We were not far into our data analysis when it became clear to us that it showed no sign of a quadrupole—none. The dipole was evident, just as we had seen it from our U-2 flights, so we felt confident that the equipment was working and that the absence of a quadrupole wasn't an illusion of our own. Once again, however, we could not be certain without first checking the southern sky. So we proceeded as planned, this time to Brazil, in October 1982. Jerry, Phil, and I were joined in Brazil by Thyrso Villela, a physics graduate student from the University of São Paulo. Our base of operations was São José dos Campos, a high-tech center of Brazil that includes one of the centers of its space program.

On the evening of November 19, 1982, our balloon was launched from the Instituto de Pesquisas Espaciais (INPE) in Cachoeira Paulista. The four of us monitored the radio signals from the balloon, which functioned flawlessly throughout the flight. Toward sunrise, the team prepared to end the flight by transmitting a signal to order the gondola to drop by parachute. To fire the bolts that secure the gondola to the balloon, one enters a series of commands by turning thumb wheels until the right number comes up, then one presses a button. When the explosive bolts are fired, someone studies the data screen to see if there's any sudden change in atmospheric pressure, a clue that the gondola is plummeting back to Earth.

At twenty-six seconds after 10:56 A.M., an INPE crew member pressed the button to trigger the drop. Nothing happened. He pressed again, and again. Still nothing. Everyone groaned. After all that work and a highly successful mission, this had to happen. The balloon was just serenely sailing off, deaf to our commands. As a backup measure, the gondola also carried an automatic timer to trigger the explosive bolts for just this kind of contingency. They failed, too.

The real tragedy was the loss of the on-board data tape. Although Jerry and I could construct a partial map of the cosmic background radiation based on the data we had been able to gather by telemetry, the tape contained important in-flight data unavailable from the telemetry—in particular, high-precision calibration data. We were heartsick as our gondola drifted onward, unreachable in the stratosphere through the late morning and afternoon.

Jerry was having mixed feelings about the unfolding debacle: At that point, with the data from the Palestine flights, he had enough data to write his doctoral thesis. He remembers thinking, I could get my thesis written faster if this thing doesn't come back. Then: God, what an awful thought! For me, I vowed that this was it—my last balloon experience. Period. A balloon lost over the Brazilian jungle—what could be worse? The uncertainties of launching the COBE satellite seemed, at that time, minor compared to the physical and mental anguish of flying these damned balloons. Never again. We gathered glumly around a table and discussed what, if anything, we could do to retrieve our wayward gondola.

As night falls and things cool, a balloon tends to contract, losing its lift, and sinks closer to Earth. In any case, ours couldn't stay up forever; its gas was slowly leaking away. The Brazilians sent an airplane to track the balloon. They managed to do so for half a day. The pilot landed to refuel, and stopped in the bar to get a drink; when he came out, it was dark. Since it wasn't a lighted field, he couldn't take off safely. He offered to do so anyway, but the INPE flight chief vetoed the idea; with darkness coming on, no one wanted to risk losing a pilot.

That night, November 20, we lost the balloon's telemetry signal. Apparently the balloon had fallen back to Earth. We scanned weather charts and maps to try to estimate where it might have come down. Judging by wind conditions, we thought it might have fallen well to the west. Fortunately, the instrument gondola carried a radio beacon. So the next morning we hired a helicopter and a fixed-wing aircraft to hunt for the balloon. The pilots flew over the jungles all day, looking for a whitish sheet of plastic draped over the trees and listening for its radio signal. They saw and heard nothing. They looked the next day, and the next. Still nothing. We knew the radio's batteries would be dead by then.

Phil went home to the United States. Jerry and I waited and searched. Still no balloon. We printed reward posters and drove around the region, nailing them to signs and telephone poles. They looked like Old West wanted posters offering a reward for information about the whereabouts of the balloon. We pinned up and handed out copies in the São Paulo airport, hoping pilots and passengers would keep an eye out. The reward was 450,000 cruzeiros—about U.S.$2,000. We would have to find the money somehow, if someone produced our payload. No one found a thing—not the slightest scrap of polyethylene. It was the Aberdeen disaster all over again, except this time we had lost entirely everything. I knew how the group must have felt when HAPPE fell into the ocean.

After six weeks, the Brazilian authorities called off the balloon search and we went home. When we got back, Phil was devastated. Several years earlier, he had been curiously chipper after the Aberdeen accident; back then he was a younger man, a graduate student, who was thrilled by the tension and mayhem of it all. Now it was 1982, he was almost thirty, and the tension and mayhem weren't fun anymore. Losing one balloon was a wild event, something a young man could chalk up to experience. Losing *two* balloons was a grind. His career was tied to this experiment. Why had he ever gotten into this crazy business? It just confirmed what I had already learned about balloons, but there I was still flying them.

We returned to California in December 1982 and assessed what we had achieved. Despite the anguish of the lost balloon, we had reason to be satisfied with the project. Our data—absent anything from the Southern Hemisphere—showed that the quadrupole report had been unfounded, which was disappointing for the two research teams who reported them. But it still left the Holy Grail to be sought.

At least as important, however, our flights had proved that cooled receivers worked. NASA officials would surely give us the go-ahead now. The cooled versions would be more sensitive than the originally proposed DMR, and enhanced sensitivity was just what we needed in order to discover the primordial seeds of cosmic structure. Cosmology had advanced a lot since the mid-1970's, when COBE was first proposed; in the interim, theorists had concluded the ripples might be much fainter than previously assumed. Hence extra sensitivity was essential.

I had to sell the benefits of cooled receivers to the other two principal investigators on COBE, John Mather and Mike Hauser, then to the overall COBE Science Working Group, and then to the engineers at Goddard. I gathered material, wrote memos, and made presentations. As I began swaying John, Mike, and the science team to my point of view, I began trying to convince and cajole the engineers at Goddard. The engineers were conservatives, reluctant to change. Based on their experience they had developed the pithy saying "Better is the enemy of good." By that they meant they had seen people change good things to make them even better, with results that were worse than the original. But I persevered.

Finally, testing my sincerity and conviction, they offered a devil's bargain: "We cannot change everything over to cooled receivers and be ready in time for the launch. If you give up one of the four DMR frequencies, we will cool two of them. We'll keep one as originally designed to ensure at least one of the original four works well." I didn't want to give up a quarter of the experiment, but they were firm. So I agreed to drop the lowest-frequency DMR, the one that was most susceptible to galactic interference. In return, they agreed to cooling the two highest-frequency channels, the ones offering the clearest view of the distant universe. NASA headquarters agreed to shift the low-frequency radiometer (19 GHz, or 1.5-centimeter wavelength) to a balloon for mapping the galactic emission. The task of flying a 19-GHz radiometer on a balloon was funded and was entrusted to the Princeton group—Dave Wilkinson, Dave Cottingham, Ed Cheng, and Steve Boughm.

Thus began the development of COBE's cooled DMR receivers, work that needed to progress in a hurry.

More than a year after we had returned empty-handed from Brazil, in January 1984, I received a phone call. The Brazilian authorities had found our balloon. I was stunned—and delighted, of course. I told Jerry and Thyrso, who had come from Brazil to Berkeley for two years to pursue his doctorate on the Brazilian flight data. Jerry and I had spent two months getting the poor-quality telemetered data from the Brazilian balloon into shape, and we were deep into analyzing it. I called Phil, who seemed dazed by the news. He had managed to survive the disaster and moved on to new research; but now the past had returned to haunt him. The National Science Foundation provided funds for an emergency trip to Brazil.

A few days later, Phil and Thyrso flew to South America. We had no idea what they would find. There was no telling how degraded the equipment might be, fourteen months after the launch. What had happened to the data tape? Even if it survived the fall, had it endured more than a year of jungle heat, humidity, rain, prying insects, and creeping fungus?

On February 3, the Brazilian team from São José dos Campos met Phil and Thyrso. They drove to a small, remote town called Tapiraí, near the site of the downed balloon. The site was more than a hundred miles from the predicted balloon landfall. The Tapiraí region was sparsely inhabited. Locals scraped by as best they could—for instance, by producing charcoal in caverns. The town of Tapiraí consisted of a few streets, shops, and hovels. Thyrso approached a few people and asked them, in his native Portuguese, if they knew anything about a fallen balloon. They did indeed; it was the talk of the town. The gondola had descended nearby. They pointed into the distance—into a forest preserve inhabited only by poachers and wild animals.

Gradually, they pieced together the story. One day in late 1983, a poacher had trespassed in the preserve. The jungle was dense, with all its eerie and cacophonous sounds. He cut down young palm trees to extract their hearts of palm; they brought a good price on the black market. He had to be quiet about it, of course; if the authorities caught him, he might face a stiff fine, perhaps even jail. And then, all of a sudden, he stumbled upon something he had never seen before. It appeared to be a shimmering shroud, very high in the trees. From the shroud hung a huge, squarish container, some tens of feet above the jungle surface. It looked metallic. He stared at it for a long time; walked around it; wondered what to do. He didn't get too close. Was it from outer space? Finally he walked away—awed, puzzled, slightly afraid, and hauling his bag stuffed with hearts of palm. He kept the secret largely to himself. He knew that if he blurted the news to anyone, the authorities might hear about it and realize he had been poaching in the preserve. Silence was smarter.

But one day he wandered into Tapirai, went to a bar, and got a little drunk. He began to chat about a strange object in the forest. At first, other bar patrons smiled; then they listened more closely. The word spread. A local cop heard about it. He traveled into the preserve and hiked around, looking for anything out of the ordinary. And there it was: a huge, whitish canopy stuck high in the trees, maybe a hundred feet up; from it dangled long cords and a gondola. He instantly recognized it as a balloon. He hurried back to town and contacted the Brazilian military. The military checked its records and realized it was the balloon missing since November 1982.

By that time, Tapiraí residents knew the object wasn't from outer space. No longer afraid, they hiked into the preserve and tugged the gondola from the trees. They broke it open and removed components. Electronic parts, shiny gadgets, whatever looked pretty or interesting—people plucked them out and proudly hauled them home. A regional newspaper ran a front-page story and photo about the exotic visitor. Somehow, the poacher heard that Phil and Thyrso were headed to town to reclaim the gondola. The poacher returned to the forest, with plans of his own.

After arriving in Tapiraí, Thyrso and Phil visited the local bar. There they saw the instrument's beam chopper (the salami slicer) and calibration mechanism—in use as a curiosity piece in the bar. Phil recounts, "We bought a few rounds for everyone and left with our chopper. We then went door-to-door looking for other parts. At the hardware store we found the chopper stand being used as a lamp, and some of the electronics had already made it into the radios of the local police."

Next, Phil and Thyrso tried to find the poacher; he wasn't around. They located some of his friends, who told them where to look for the payload. A local resident had a truck and offered to drive them there. On the morning of February 6, with a group of seven, they drove into the preserve. It had been raining heavily, and the roads were pure mud. Sometimes the truck started skidding, and everyone got out to push it. Mud spewed everywhere, the truck lurched forward, and they clambered back aboard, only to get stuck again later.

Finally they reached "literally the end of the road." They saw the poacher's hut; he wasn't home. They abandoned the mud-covered truck and hiked into the forest. The hike was leisurely, save for the locals' relish in telling stories about "the large snakes that have been known to eat people." Then they saw the white balloon plastic and the orange and white parachute canopy, spread over distant trees. As they approached, they looked for the gondola. It was gone. In its place was the poacher, who was calmly seated on the ground and whittling a palm branch.

What followed was a high-speed conversation in Portuguese. Phil stood by, baffled by it all. The poacher explained to Thyrso that his friends in town had warned him the scientists were coming to reclaim their flying machine. He didn't want to give it up—and why should he? He had found it, hadn't he? And since he was being asked to give up personal property, didn't he deserve compensation? Phil began to realize what was going on. The poacher said he had hidden the gondola and wanted to negotiate a deal for its return. To be specific, he wanted three things. First, three thousand dollars in U.S. money. This was many times the per capita annual income in Brazil. Second, a television interview. Third, an interview in a major Brazilian newspaper. Evidently, the poacher had lost his fear of preserve authorities.

Phil replied, via Thyrso's translation, that the first demand was unacceptable, and the second and third were beyond their control. More negotiation in rapid Portuguese ensued. The poacher sighed, turned, and left for his hut. He was calling their bluff. Phil was furious.

Meanwhile, one of the group noticed a path in the forest and decided to walk down it. About a mile down the path, he discovered the gondola. Angry, but happy to get the gondola, Phil paid the poacher about a hundred dollars and promised to get his name in a U.S. newspaper, which we did—in an article in the local UC Berkeley paper. In any case, it was a good deal less than the two-thousand-dollar reward we had originally offered, so Phil could look upon the transaction as a bargain.

Ultimately, we recovered every major component of the instrument, although after fourteen months in the forest, some were rather corroded. Amazingly, the dewar was intact and still retained vacuum; it is currently being reused. The most important item was the on-board cassette tape, which contained important data that we could not reconstruct from the telemetry, in particular a very precise instrument calibration that was unique to this flight. Also amazingly, although the tape was covered with a green fungus, we managed to read it into the computer. The data were 98 percent usable, and they confirmed what we had found in the Northern Hemisphere: no quadrupole.

* * *

Clearly, cosmic anisotropy—the wrinkles of creation—was proving far harder to detect than most of us had ever suspected. In the late 1960's, theorists had suggested anisotropy might be relatively easy to spot—perhaps as obvious as one part in ten of the cosmic background radiation. But by the mid-1980's, our instruments were nearly as sensitive as one part in ten thousand, and still the seeds eluded us. We were fast approaching the limits of resolution possible within Earth's atmosphere. If the seeds really existed—if they weren't just some awful theoretical error—then our best hope was to seek them from space, with the COBE satellite.

NASA had accepted our proposal with one significant change. We had recommended flying COBE aboard a Delta rocket, but by the early 1980's, NASA was trying to phase out such expendable vehicles and shift all launch business to a new type of spacecraft: the manned space shuttle. Though a manned mission meant more care was needed for safety, the shuttle looked like our path to space and we were happy to be riding it.

Chapter 11

COBE: The Aftermath

Tuesday, January 28, 1986: Most people remember where they were that day. I do.

It was to be a banner morning in the history of the U.S. space program, for two reasons. At the Jet Propulsion Laboratory in Pasadena, California, scientists were preparing for *Voyager 2*'s successful flyby of Uranus—the most distant planet ever reached. And on the East Coast, Christa McAuliffe, the first schoolteacher to join a space crew, was sitting atop the space shuttle *Challenger* at Launchpad 39B at Kennedy Space Center, near Orlando and Titusville, Florida. With her were six other crew members, including an African-American, an Asian, and another woman: a crew of unusual ethnic and sexual diversity. The future looked bright for the shuttle. Despite its technical limitations and financial shortcomings, it had already fulfilled its promise as a satellite "repair shop" in space. Within a few years it was to be used to assemble the space station, and perhaps even larger structures: manned spaceships to the Moon? Mars? Who could guess what the future might hold? Space travel had come to seem almost routine, "almost like taking the E train." Two members of Congress had flown on the shuttle; Walter Cronkite might be next. The cosmic background explorer (COBE) satellite was scheduled as a payload in a little more than two years' time.

Like much of the nation, I planned to watch the launch on television. It had been scheduled for 9:00 A.M. eastern standard time, which is 6:00 A.M. in Berkeley, on the West Coast. I got up early to watch, and began getting ready for the day, keeping an eye on the countdown. The shuttle had flown more than twenty times, and so this launch should have been routine. It was special only because of the crew's novel composition. The morning was unusually cold at Kennedy Space Center, with frost forming in places. Some engineers counseled against launch, warning that the temperature was outside the safe operational range. Under pressure to maintain the shuttle's schedule, especially with the glare of publicity about the schoolteacher McAuliffe, NASA officials decided to launch anyway. There would, however, be a delay. I decided I couldn't wait around any longer and set off for work about 8:30 (11:30 EST).

At 11:37:53 A.M. eastern standard time, the countdown ended with ". . . three . . . two . . . one . . . zero—ignition!" No one could see—or would see until later, on closeup videotapes of the igniting engines—a small, dark puff of smoke that squirted from one of the solid fuel rockets. The smoke soon became a small flame. The shuttle gushed fire and smoke, and slowly lifted from the launchpad. The crowd gasped with excitement. At an ever-accelerating rate, the craft arced into the sky, a white trail marking its route. Sixty-seven seconds into the launch, the shuttle, now barely visible, was briefly engulfed in an orange-white ball of smoke, and white trailers began to arc back toward the Earth. The crowd, not knowing exactly what to expect, applauded hesitantly, thinking the occurrence to be a stage in the ascent. The solid booster then broke free and began spinning; the hydrogen container in the external tank blew open, releasing its twenty-five thousand gallons of liquid hydrogen; the solid booster hit the part of the external tank that housed liquid oxygen and burst it like a balloon filled with water. The liquid hydrogen and liquid oxygen mixed; the mixture detonated.

Challenger had exploded catastrophically.

The trip from my house up the hill to the lab takes twenty minutes, time for a few thoughts about the day's work. One of my tasks that day was to check through the calibration data of the differential microwave radiometer. It was part of the unfolding program to have our instrument ready for the COBE launch on the shuttle, which, according to NASA's most recent estimate, would be late in 1988. Members of the differential microwave radiometer instrument team, motivated by the scientific goals and engineering challenge, were just beginning to meld as a unit. There were still some rough edges and some difficult challenges ahead. If I could get them through testing the first of the three radiometers, then things would slip nicely into place. I reflected on the crowd waiting for the *Challenger* launch that day, the last thing I'd seen on the TV as I left the house. One day, I mused, we would be part of such a crowd, watching as fifteen years of scientific labor came to fruition.

On my arrival at the lab, I knew something unusual had happened, as groups of people were huddled in corridors, in offices, around radios and a couple of television sets. In the short time between leaving my house and arriving at work, seven astronauts had been killed.

I was stunned. We all were. We grieved for the astronauts. The tragedy of the accident was uppermost, but slowly the probable implications for COBE began to dawn. Our project had operated on one key assumption: It would fly aboard the space shuttle. But now, NASA was likely to suspend shuttle flights, for a while at least. I have to admit that, perhaps naively, I thought the delay might not be very long. But I was soon disabused of that fantasy. NASA indefinitely delayed all shuttle flights while accident investigators analyzed footage of the explosion, engineers pored over telemetry records, and navy crews probed the ocean for shuttle fragments.

With one shuttle lost and three grounded, NASA's schedule had gone to hell. Nothing was flying. There was no telling how long the COBE launch might be delayed; maybe years. It could get worse. Before the accident, the shuttle wasn't flying nearly as often as originally planned, and customers, both civilian and military, were standing in line seething over delays. Once shuttle flights resumed—and we had no idea when that would be—there would be many people further up the line than we were. Officials with high-priority military satellites and megascience, megabucks projects like *Freedom, Galileo,* and the Hubble Space Telescope would be pushing and shoving to board the shuttle. COBE's launch might be delayed well into the 1990's, when getting a shuttle ride could become even harder. By that time, the shuttle would be needed for numerous, closely timed missions to ferry astronauts to orbit where they would construct the *Freedom* space station. It was all too clear that COBE—a medium-sized project, with few political connections and little public visibility—could get lost in a frenzy of rescheduling.

While most of us were wrestling with how the shuttle disaster would affect COBE, Dennis McCarthy, COBE deputy project manager, assessed the situation. The day after the shuttle explosion, he called a COBE project meeting. Gloom marked every face at the table. He stated bluntly: "I don't think we'll ever launch on a shuttle from the West Coast. We've got to have a different way of launching this thing." Something in Dennis's demeanor stirred us, shook us out of our dazed inertia, and made us begin to believe that, yes, there might be a way around this awful thing. People began to suggest ideas and set to work. Although we did not know it at the time, this was the beginning of the most intense, and most rewarding period of the COBE project. The enormous odds we faced galvanized our team spirit. The *Challenger* accident—great tragedy that it was—energized the COBE team in a quite extraordinary manner.

We needed to find another rocket—and quickly, because few were available. Just six Titan 34Ds, three Atlas-Centaurs, thirteen Atlases, and three Deltas remained, and other space customers might soon snap them up. We couldn't just pick any rocket; we had to pick one that was compatible with COBE's size and mission, or one that could accept a redesigned COBE. We needed a plan for redesigning the spacecraft, and for selling it to the management, both at Goddard and at NASA headquarters. We established only two ground rules. First and foremost, science objectives could not be compromised, if at all possible and practical. Second, and equally important, we would attempt to maintain the same launch schedule; for every week that we ran past the deadline we would incur extra costs, and that would make our approach difficult to sell. It was to be a year of never-ending, sometimes desperate meetings.

Dennis's insistence that we find a rocket for launching COBE was soon justified. Three months after the *Challenger* accident, there were persistent rumors that the air force planned to stop construction of its southern California shuttle launch facility at Vandenberg Air Force Base. Our scientific goals required a polar orbit and we could only achieve this by launching from California. If the air force carried out its plan (it did), we would be a satellite without a spaceship. But lack of a shuttle launch facility at Vandenberg eventually became an academic problem. In 1986, NASA announced that, whatever happened, COBE would not get a place on the civilian shuttle when it resumed flying. Thanks to Dennis's push start, we already were looking for alternatives.

Amidst growing despair, I had one thing to look forward to: a South American trip and Pacific cruise. In March 1986, Halley's Comet was making its seventy-six-year, periodic return to the Earth's vicinity. I had agreed to lecture on an adventure trip to Ecuador and "comet cruise" around Darwin's Galápagos Islands. Guests had paid thousands of dollars for tickets.

Comets, unfortunately, don't always live up to their billing, and Halley turned out to be extremely dim on this pass around the Sun. On the trip, no one seemed to care; the whole voyage was so romantic, the stars so breathtaking, and (I hope) the lectures so interesting that they had a wonderful trip despite Halley's disappointing performance. Night after night everyone went out and looked at the comet, the Magellanic Clouds, and the spectacular southern sky. Their enthusiasm refreshed and invigorated me. I thought: Look at these people who are so thrilled, they'll pay thousands of dollars to learn about what's happening in the sky. Yet I work on it every day. I realized how privileged I was, and that knowledge boosted my spirits as I squinted to see Halley, a barely perceptible glow over the dark waves.

The comet's name honors the astronomer Edmund Halley, who anticipated key features of modern cosmology, such as Olbers's paradox (which Olbers "discovered" eighty-four years after Halley's death in 1742). Halley also found the first direct evidence that the "fixed stars" are not fixed but, in fact, are in motion. This discovery was a major step away from the old idea of a static cosmos and toward the modern concept of a dynamic, evolving one. Halley's greatest contribution, however, was to use Newton's celestial mechanics to determine the comet's orbit and predict its return in 1758. Return it did, sixteen years after Halley was dead and buried. Europe in the age of Enlightenment was stunned: Once again, human reason had triumphed anticipating nature.

Halley's prophecy was the first spectacular vindication of Newton's theory of gravity, one that could be appreciated by any literate person. The pseudonymous "Astrophilus" boasted in *Gentleman's Magazine* at the time: "I cannot but congratulate my countrymen on an event so glorious to the Newtonian doctrine of gravity, and to the memory of that excellent philosopher Dr. Halley; and may it ever be remembered that the first instance of an event of this kind was foretold, and with accuracy too, by an Englishman." Wrote another: "By [the comet's] appearance at this time, the truth of the Newtonian Theory of the solar system is demonstrated to the conviction of the whole world, and the credit of astronomers is fully established and raised far above all the wit and sneers of ignorant men." True, comets would remain mysterious well into our century; in Edmund Halley's day, some writers speculated comets were the abode of damned souls, and as late as 1910 many feared the comet's tail would poison Earth's atmosphere. Even so, Halley's one great prediction established once and, seemingly, for all that gravity ruled the heavens.

But did it still rule in 1986? Few cosmologists could look at Margaret Geller and John Huchra's recently published redshift maps of giant galactic superclusters without wondering if gravity had, finally, met its match. Could gravity alone have formed these massive structures in the 15 billion years since the big bang? Perhaps some other, unknown factor was needed to explain the origin of such material extravagance. If we ever got it launched, COBE might provide an answer. The detection of temperature variations in the cosmic background radiation—the imprint of cosmic seeds present three hundred thousand years after the big bang—would confirm that gravity molded the present universe. If we failed to detect the seeds, we would have to seek other explanations.

By the end of the cruise, and the comet's timely appearance, I was reenergized and ready to get back to the lab.

COBE was still a satellite without a way to space. The COBE team was struggling to locate a rocket that could accommodate it. Of those available—the Delta, the Atlas-Centaur, and various types of Titan missiles—the Atlas-Centaur was ruled out quickly because Vandenberg didn't have a launch pad suited for these rockets. The Delta would work, but it would require draconian redesign of the COBE spacecraft, reducing its size and weight by half. The Titan looked more promising; its payload area was spacious enough to hold the current COBE with only minor modifications to the satellite. A drawback was cost—as much as $250 million.

The fates have a strange sense of humor, as we all discovered when we opened our April 19 *New York Times* and saw the front-page headline: TITAN ROCKET EXPLODES OVER CALIFORNIA AIR BASE.

A Titan 34D had been launched from Vandenberg and, within seconds, burst into a large orange fireball. It spewed noxious gas over hundreds of acres. At least fifty-eight people were treated for skin and eye irritations; schoolchildren stayed in classrooms until the vapor passed. It also destroyed the launchpad. That event put a pall over the Titan prospect for COBE.

Two weeks later, a 116-foot-tall Delta rocket carrying the GOES-G weather satellite was poised on the launchpad in Florida. It was to be NASA's first launch since the *Challenger* and a bid to regain public confidence, to show that NASA still had the Right Stuff. "We need to remind ourselves that we have had success in the space program," NASA chief William Graham said. The Delta soared skyward. Seventy seconds into flight, it lost power in its main engine and pitched out of control. Lest it fall on inhabited areas or ships at sea, a ground controller pressed a button to destroy it in flight. Again *The New York Times*'s front page: THIRD U.S. ROCKET FAILS, DISRUPTING PROGRAM IN SPACE.

The aerospace industry was "in chaos," a top McDonnell-Douglas official told United Press International. "It's confused; there's just no launch capability that exists now and it's going to be that way for several years." Several years! That was the last thing we needed to hear while we were struggling to save COBE.

The Delta blast was especially perplexing; it had been considered among the most reliable rockets in the U.S. inventory. Ugly rumors circulated; some speculated (according to *Science* magazine) about the possibility of sabotage via "a radioed command from an external source."

We had contemplated turning to the French Ariane rocket, the mainstay of the European Space Agency. Ariane had built a reputation for being a trustworthy vehicle, despite some very public accidents. It is launched from French Guiana, so COBE would have to be shipped there—a drawback, but not fatal to the aims of the project. "We had two or three discussions with the French," one COBE team member recalls. "But when [NASA] headquarters found out about it, they ordered us to cease and desist—and threatened us with bodily harm if we didn't." U.S. pride was at stake. In the wake of so much bad publicity, NASA didn't want a major research project running to the French for help. We were given no promises, but had renewed hope we could achieve a U.S. launch.

The most exciting possibility—the dream COBE mission—was to launch the satellite to a much higher altitude than originally planned. In the void around the Earth and Moon are several places known as Lagrangian points, after the eighteenth-century astronomer who discovered them.* At these points, the Sun's, Earth's, and Moon's gravity balance, so that an object would orbit there, suspended, like a Christmas display on a cord between two buildings. One point, known as Lagrangian point 2 or L2, would be especially good for a cosmological satellite because it would receive virtually no radiation or electromagnetic interference from Earth or the Moon; also, if put in orbit around the L2 point, it would constantly be in the sunlight instead of passing through occasional eclipses, which could briefly diminish power to its solar cells.

Conceivably, if we could cajole NASA into manifesting COBE on a shuttle flight, then the mission might go like this: The shuttle could launch from Florida into a regular equatorial

*To the space-minded public, the best-known Lagrangian point is L5. In the 1970's, Princeton particle physicist Gerard O'Neill advocated L5 as a future site for the construction of large space colonies.

orbit. Then, using a new upper-stage rocket developed by Orbital Sciences Corporation, COBE could be propelled to L2. We loved the idea. COBE project notes report that when the L2 idea was broached, all the principal investigators "acted as if they had had a religious experience. . . . It was the most benign possible environment [for COBE]. . . . It really made everything easy." The Soviets had also talked about the possibility of sending a cosmological satellite to Lagrangian point 2. Unfortunately, the upper stage would have cost a couple of hundred million dollars. So that was the end of that dream.

With limited real choice available, in late 1986 the COBE project recommended that NASA launch the satellite aboard a Delta rocket. Despite its May 3 accident, the Delta program still had a superb flight record—May 3 was its first failure in nine years. Also, the Delta was a relatively cheap rocket, so NASA headquarters would be happy.

McDonnell-Douglas built the Delta rockets, but the production line had been shut down, supplanted by the shuttle program. Enough parts were left to make a few more Deltas, and these had been abandoned in old sheds. The "Star Wars" program immediately bid for them, to be used as targets for antimissile weapons. We fought for one—successfully. McDonnell-Douglas engineers practically had tears in their eyes when they learned they would get a chance to build one for us. They had spent years building the rockets, often for military uses. They didn't want to see the last Delta rockets being smacked together in antimissile tests. They preferred that their rocketry talents go to scientific research of lasting merit—like COBE.

Delta, however, was a small rocket. It could loft a payload only half as large as the existing COBE-shuttle configuration. Hence, we would have to make COBE much smaller and lighter. This was even more difficult than it may sound, for we couldn't simply make everything smaller, like a Russian doll that contains a smaller version of itself. The smaller COBE became, the more its instruments might begin to interfere with each other, like a family that moves from a five-room to a one-room house. The three instruments were already highly sensitive to stray heat sources, electromagnetic radiation generated by wires and other instruments, and so on; their vulnerabilities would grow as COBE shrank.

On October 1, 1986, we received news: NASA headquarters gave the go-ahead for a Delta launch in early 1989. That was much sooner than we had expected. "As soon as they said the launch date, everybody knew it was going to be a fire drill . . . an incredible schedule," said Frank Kirchman, theDMR's new thermal engineer. From that point on, we were committed. It was "back to the drawing board," or nothing.

To transform COBE so radically so fast, the teams would have to be as tight and intimate as possible. Roger Mattson, COBE project manager, chose to go with a "skunkworks"-type operation, that is, one that was "under one roof." There would be "no limit on overtime and/or compensatory time, as necessary to maintain schedule," one memo records. Pre-*Challenger* COBE teams that had been miles apart would now be literally down the hall from each other, a move that would speed up the work and make people more accountable: People are always more accountable when you can look them in the eye instead of dealing with their answering machines. The clock was ticking.

The three COBE instruments—DMR, FIRAS, and DIRBE—were built separately, in different buildings. For the DMR effort, which was centered mainly in Building 19 at Goddard, I was the principal investigator and Chuck Bennett was deputy; John Mather and Richard Shafer were principal investigator and deputy for FIRAS, respectively; Mike Hauser and Tom Kelsall served these roles for DIRBE. The overall manager of instrument development was Earle Young, who faced a basketful of headaches in those years.

The DMR work area in Building 19 was junky looking and jokingly called the Microwave Instrument and Venetian Blind shop. There the engineers built the parts and tested the components of the DMR. They assembled the DMR and test equipment. However, they mainly worked on the calibration and testing necessary to make the DMR ready. "To say this *thing* was 'designed' is a little bit charitable," remarked Roger Ratliff, a master mechanical engineer who made many of the DMR's parts. It wasn't Rube Goldberg, but we certainly operated by the maxim of "whatever works, works."

NASA threw its resources behind us after realizing how COBE personnel were fighting to keep the project alive. NASA's goal was to launch a space mission to show the public that NASA wasn't down for the count, that it was still in the space business and still capable of pulling off an impressive scientific achievement. Word got around Goddard that NASA considered COBE a "sexy" mission, one that deserved to be saved: We heard that "COBE was a project that was very interesting to the public, and it was very interesting to the Congress"; that "it was high-visibility—you know, the big bang! The origins of the universe and all that! And there would be a lot of interest in the results if it came out all right." That kind of talk makes a big difference in team morale: "I knew it'd be pretty hot stuff," said Roger Ratliff, one of NASA's "old guard." Rob Chalmers, the spacecraft thermal engineer, agreed: "If you can figure out some way to focus the troops, they will work themselves almost to death for you."

The scope of our task was crystallized in the Goddard *Engineering Newsletter:* "The transformation of COBE from an STS [space transportation system, a.k.a. space shuttle] to a Delta launch is probably one of the greatest engineering challenges ever undertaken by the GSFC [Goddard Space Flight Center]." With that kind of challenge a good team can respond magnificently. Ours did. Morale soared.

Then, barely a year into our frenzied program, a major development in cosmic background radiation research stunned the world of cosmology. Once again it reminded us that we might very easily be beaten to our goal. We held our collective breath as the news came through.

In February 1987 a joint Japanese-American team had launched a suborbital rocket, carrying a detector built by Andrew Lange, Paul Richards, and their colleagues at Berkeley. The Japanese team was led by Professor Matsumoto of Nagoya University. During its brief flight in space, up to 320 kilometers high off the Japanese island of Kyushu, the rocket-borne instrument analyzed the background radiation at six frequencies between 0.1 and 1 millimeter. The result, which became public in August that year, showed a dramatic difference between the predicted and measured curves of the cosmic background radiation spectrum, with an excess temperature, rising as high as 3.18°K. If real, it meant some extraordinarily energetic events had transpired in the early cosmos.

The excess caused tremendous excitement. It provoked theorists to offer several possible explanations. The most far-reaching challenged the standard big bang model, which required the spectrum of the background radiation to conform nearly perfectly to blackbody radiation, and yet the new result seemed to indicate it might be significantly distorted. This possibility added to a growing unease about the validity of big bang theory. A report of the Nagoya-Berkeley project was included in an acclaimed Public Broadcasting television series, *The Astronomers.* Only after the filming was essentially complete did it transpire that the reported distortion in the spectrum of the cosmic background radiation was incorrect. Almost certainly, the excess was an illusion caused by instrument design or malfunction, or a false signal caused by extraneous radiation or noise.

We were chastened by the Japanese-American experiment. It reminded us that at any time we might hear of another result, from another team, but this time announcing the discovery of ripples—no mistakes, the real thing. It added an urgency to our work, and suddenly NASA's early-1989 launch date, which had once seemed unrealistically early, now looked distressingly late. We worked even harder, tackling scores of problems: vibration, systematic errors, the potential effects of residual oxygen on the mirrors, stray sources of heat—they all had to be resolved, and they all consumed time, expertise, and emotional energy.

In the original COBE blueprint, the differential microwave radiometers would stand on pedestals around the liquid helium dewar, which housed the FIRAS and DIRBE instruments. A skunkworks team led by mechanical engineer Gene Gochar developed a new DMR design. The radiometer housing was made wider and shallower. The lower part of the DMR had to curve

NASA official Len Fisk (*left*) inspects the 53-GHz DMR. Engineer Rick Mills explains.

around part of the dewar because space was so tight. Wiring, thermal design, and other factors were also revamped as part of the COBE redesign.

A major task was the replacement of the old, stiff thermal shield with one that was lighter and flexible, which would unfold in orbit. The task of building a flexible yet reliable shield "was viewed by all as extremely challenging, if not impossible," Rob Chalmers wrote in the Goddard *Engineering Newsletter*. Yet they pulled it off. Also, the solar cells that would have been wrapped around old COBE's outer wall were replaced with cells on three accordion-like panels; these, too, would be unfolded in space. The new COBE also had to have a smaller antenna for sending and receiving data and commands. These changes were difficult and posed many worries. The next couple of years were both interesting and anxiety-ridden.

Once we finished rebuilding the radiometers, we had to test and calibrate them. This would be the ultimate test of our instrument—and of our senses of humor. The trick to calibration is to simulate the background radiation. This is not easy. The background radiation is extremely faint, and so we had to simulate it with a "target" of similar faintness—that is, one with a temperature near absolute zero. Then we had to point the two antennas of the radiometer at targets with essentially the same temperature. If the antennas were working perfectly, they would both detect the same temperature and, therefore, report a temperature difference of zero. If the radiometer detected a temperature difference, then something was wrong: that is, the radiometer must be sensing a stray, undesirable source of radiation (for example, an internal heat source, a magnetic field, a radio source, you name it). We first calibrated the radiometer at Goddard in a large room lined with dark, cone-shaped objects that absorb microwave radiation. They also tend to absorb sound. When you spoke in the room, your words seemed to stop just outside your mouth.

To simulate the vacuum of space, we also "remotely" tested the differential microwave radiometer inside an evacuated container. The container was forty inches wide and several feet tall, and located at the rear of the "clean room," a special facility that was essentially dust-free. The DMR at each frequency was about two feet across, a foot thick, two feet tall, and weighed roughly one hundred pounds, so we could test only one radiometer at a time. Inside the evacuated container, a target sat on a track, like a toy train; we controlled its motion from the outside, moving it back and forth in front of the radiometer horns. The DMR normally looked at a spinning wheel covered with microwave-absorbing cones, each of which was supercooled from the rear (via tubes connected to a dewar of liquid nitrogen). The spinning would, we hoped, "even out" any warm spots on the wheel and thereby mimic the uniform background radiation even more accurately than a static target.

This clever gadget broke down annoyingly often. Parts stopped moving because the extreme cold froze grease or other lubricants, or shrank components so that they no longer interacted as desired. Each time the system broke down (inevitably, in the middle of someone's all-night shift), it had to be slowly rewarmed, then refilled with air so that we could get in there to see what was wrong. Rewarming could take hours or days; you don't want to do it too fast lest delicate components be damaged by the sudden rise in temperature. The frustrating thing was that just before we finished warming the chamber and were ready to reopen it, whatever was broken—for example, the target on its track—would start working again. Hence we had no way of knowing what had caused it to break down in the first place. We would then cool it down for another round of tests, only to have it break down again. We couldn't see what was going inside when things went awry because the chamber was sealed tight, with no windows (for a good reason: to prevent stray radiation from

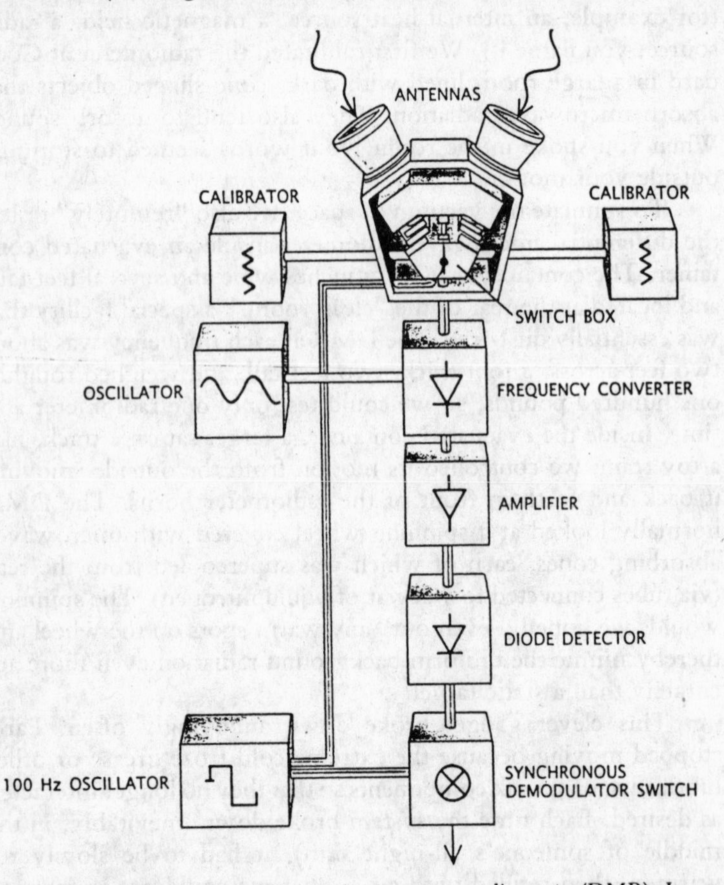

The configuration of the differential microwave radiometer (DMR). It measures the difference between the microwave radiation received from the two points on the sky with two horn antennas that are alternately connected to a single receiver. NASA/GSSC

leaking in). We considered placing a TV camera in the chamber, but there was not enough room. For months, the DMR lab was the scene of much gnashing of teeth.

In the process, we all worked very long hours, often over seven-day weeks that dragged on for months. Sleeping habits suffered; so did personal lives. Despite the stress, people managed to keep their sense of excitement. We had the feeling we were doing something deeply important. Roger Ratliff put it this way: "If it hadn't been so much fun, I don't see how anybody would have had the patience to work through it." As tense as we sometimes grew, as angry as we occasionally became with our instruments or with each other, we were united by the knowledge that our work might yield historic scientific pay dirt. We felt like a team, one that, in my career, has been unequaled in spirit and determination.

As team leader I had worked hard to forge a dedicated and cohesive team by involving team members deeply in the overall goal and by giving them as much responsibility as possible (something I'd learned from my former group leader, Luie Alvarez). Each person was responsible for and got credit for a specific part of the effort. As a result, people often became quite proprietary about their part of the instrument.

For instance, Maria Lecha was the lead engineer on the 53-GHz radiometer and was so protective of the instrument that she called it her "baby." Maria was in her mid-twenties at the time, with a degree in electrical engineering from the University of Puerto Rico in Mayagüez. She was obsessed with the project; once, after a snowfall, she went outside and made a "snowman" in the shape of the DMR. During the months of instrument testing, she spent many late nights at her Greenbelt home in front of her computer, remotely monitoring the status of her radiometer. A number of times she noticed its temperature was rising—the cooling system had died. So she'd hurry to work in the middle of the night—the Goddard guard waved her through—to rescue her radiometer. "Millions of dollars were in my hands," she recalls. "I was always afraid something would go wrong, that it would explode on the rocket during takeoff." She was not alone. Bobby Patschke, lead engineer for the 31-GHz radiometer, confessed once, "I dream about my radiometer."

* * *

We hoped to ship COBE by late 1988 to California, where it would be merged with the Delta rocket. Tensions ran high as the sand ran through the hourglass. "We had some goddamn screaming matches between the engineering side and the science side. I'm not kidding," Tom Kelsall, DIRBE deputy principal investigator, recalls. "It sometimes got emotional, but it was never personal. It's like they say in the Mafia—it's business, nothing personal!" Despite our conflicts, I've got to hand it to the engineers: In the last analysis, they were on our side. Dennis McCarthy told his colleagues: "Hey, guys, we're building this thing for the scientists—not for us. And we've got to make sure they get the data they want." Agreement came from other engineering staff: "Perhaps more important for many of us was that we didn't want to let guys like Mike Hauser, John Mather, and George Smoot [the COBE principal investigators] down," Mike Ryschkewitsch said.

Referring to Hauser, Mather, and myself, one COBE project member said: "These three guys were constantly on our backs about things that could screw up their data. It never got through my fat head what they meant until I saw that show *The Astronomers* and saw the thing about the Berkeley-Japan [CBR] experiment that made the mistake. Then I thought, Thank God those guys were on our backs all along!" Whatever had happened to the Berkeley-Japan venture, we didn't want *our* expedition to fail because of some design fault or engineering mistake. If we failed to find the ripples for some legitimate scientific reason, okay. No other reason would be acceptable to us. Too much rested on the project.

Finally, we had achieved our goal: The instruments had been reconfigured, and COBE was half its previous weight. The construction of COBE had involved more than a thousand people, consumed many years of effort, and cost some $160 million. We were ready to ship the satellite to Vandenberg Air Force Base for its rendezvous with the Delta rocket and its eventual launch.

Chapter 12

First Glimpse of Wrinkles

The launch of the cosmic background explorer satellite was scheduled for November 18, 1989. I drove the three hundred miles south from Berkeley to Vandenberg Air Force Base the day beforehand. The rolling, amber-brown California hills, carved by millions of years of plate tectonics, earthquakes, and erosion, had the look of autumn about them. Soon we would be facing our moment of truth. I had been in cosmology almost two decades, and for most of the time I had been working toward this moment. The three instruments aboard COBE were going to explore the early history of the universe in a way never before possible. Many believed that COBE was going to resolve once and for all whether the big bang really happened. For me, there was no question about it: The big bang did happen, and we were going to find out much more about how it happened.

My mind was strangely relaxed. COBE's fate was in the hands of the rocket engineers. All I could do was watch it go up—or blow up.

Vandenberg is a desolate place on the California coast and, in many ways, a dangerous one. Fog blankets the area frequently, and motorists are warned to drive carefully. Cliffs are unstable ("Do not venture to the edge of a cliff . . . it may give way," an air force brochure cautions visitors); the thirty-five-odd miles of beaches include dangerous riptides; unexploded bombs are scattered across the terrain ("If you see a suspicious object, DO NOT attempt to handle it or take it as a 'souvenir'!"). Drivers sometimes screech to a halt as a wild pig or deer scampers out of the fog. The 98,400-acre base, which served as an army outpost and POW camp in World War II, was named for former Air Force Chief of Staff Hoyt S. Vandenberg. For more than three decades it has been the launch site for almost two thousand rockets, including U.S. ICBMs that were propelled into space or to targets thousands of miles across the Pacific.

The nearby community of Lompoc is mostly an air force town, the kind seen in old movies: little shops, tree-lined streets, and friendly garages where they still offer to pump your gas. I drove through town, then took the northwest road through barren countryside to Vandenberg. There, as I rounded the bend, was the Delta 5920 rocket—erect on her pad, snug within the steel girders that would shield her from the elements until shortly before launch. The Pacific glistened about a half-mile away, at the foot of a gentle slope; you could smell the ocean and just barely hear the rolling surf. The salt from the sea is corrosive, so much so that the metal gantry, which supports rockets prior to launch, has to be cleaned regularly to prevent damage. One Vandenberg official joked, "Park your car and leave it here for a month—you'll see what it's like."

Thirty years earlier, another Delta rocket had changed cosmological history. But that first time, no one had expected it to do so. On August 12, 1959, NASA's first Delta rocket launched the *Echo 1* satellite, the giant spherical balloon that began the age of communications satellites. In order to "hear" signals bounced off that satellite, Bell Laboratories had built its horn-shaped microwave antenna in Holmdel, New Jersey—the one with which Arno Penzias and Robert Wilson serendipitously discovered the cosmic background radiation, the residual glow of the fury of creation, an echo of the distant big bang. Now we hoped COBE would continue the revolution that began in Holmdel a quarter of a century ago. "We're like space paleontologists, searching for the equivalent of dinosaur bones in the early moments of the universe," I told one reporter.

Before the *Challenger* disaster, this launch facility had looked doomed. The shuttle appeared destined to inherit the launch industry, lock, stock, and barrel, and McDonnell Douglas officials were thinking of abandoning the site. But, according to McDonnell officials at the base, their local manager—Pat Conlan, who has since died—fought successfully to keep it open. "To him, the idea that the shuttle would haul everything was 'patently stupid,' " one company official recalled. "If he hadn't been so determined to keep it open, McDonnell Douglas might have left the place. Period." And we would have had nowhere to launch COBE.

The temporary abandonment caused a problem. The launch gantry, known as SLC-2W, was, to quote a memo, "unbelievably DIRTY!!!" COBE had to remain as clean as possible to prevent contaminants from damaging the instruments or confusing them in orbit. Vandenberg officials agreed to clean the gantry thoroughly before COBE was installed in the Delta. Anticontaminant experts carefully inspected the gantry with "black" (ultraviolet) lights and "tape lifts" (in the latter case, technicians lay sticky tape over objects, then, using a microscope, scan it for contaminants).

The Delta rocket arrived several weeks before launch. It was shipped by truck from Florida, where it had undergone prelaunch assembly and testing. At that time, the Delta crew was down to a single team and they were also busy preparing a launch from Kennedy. A few weeks before launch, COBE began its long journey. I felt elation and sadness as I watched COBE leave its isolation clean room and be loaded into its special shipping container for the trip west. It was flown from Andrews Air Force Base directly to Vandenberg. After arriving, COBE was put in a special clean room where the final flight pieces were assembled. I made a second special trip to Vandenberg to inspect the Sun shade and Earth screen. We tested everything one final time.

Then, after midnight, when traffic was minimal, COBE was hauled in a transport vehicle to the launch facility at Vandenberg. The vehicle covered the eleven-mile journey at just a few miles per hour, so it was a slow trip. Security personnel helped with the satellite transfer by following along, their lights flashing; as one Goddard engineer observed, "You don't want people going through intersections when you're hauling something that precious."

As the launch neared, we figured we had guarded against every contingency. We had—except for one that hadn't crossed our minds. Dennis McCarthy was watching the World Series on TV and saw an earthquake shake San Francisco, three hundred miles up the California coast. Agitated, he called to Vandenberg to find out if COBE was okay. They checked—it was. As I lived in the San Francisco Bay Area, I had more immediate problems on my mind. The 7.1-Richter quake had wrecked parts of the Bay Area and killed more than sixty people. I quickly checked around work and determined that everyone in my group was fine and the equipment was okay. It was a disturbingly close call—a reminder that in life, the really big disasters are often the ones you least expect.

On Friday, November 17, NASA held a prelaunch press conference in Lompoc. John Mather, I, and many Goddard and NASA staffers were there. So were Ralph Alpher and Robert Herman, George Gamow's colleagues who, forty-one years earlier, had predicted the existence of cosmic background radiation. They looked happy to have lived long enough to see their work bear such elegant fruit. COBE, Alpher said, "is a marvelous, sophisticated instrument that we hope will provide the most accurate information we have ever taken" on cosmic origins. The Delta also came in for praise: "the world's finest launch vehicle," NASA official Don Tutwiler called it. Reporters were told that the launch "window" was from 6:24 A.M. to 6:54 A.M. the next morning; that the two-stage, 116-foot-tall rocket had a liquid-fueled first stage with nine strap-on solid rocket boosters; that the rocket, at launch, weighed more than 440,000 pounds; and that the first stage had been built by McDonnell, the second by AeroJet, and the solid boosters by Thiokol Corporation. To attain escape velocity, the rocket must fly faster than 18,000 miles per hour. In orbit, it would eject the satellite, which would circle Earth at about 16,800 miles per hour.

NASA official Larry Caroff said the satellite and other planned space observatories "are going to provide us with an unprecedented, unbelievable view of the universe. We will see things we haven't even imagined. This is going to shake up the world of science." John Mather cautioned reporters: "I am not expecting to overturn the big bang theory with what we see, because it is a good theory and works well. However, we could get a big surprise."

Saturday morning. I rose early, showered, and dressed warmly. In the predawn darkness, not far away, fifteen years of work were sitting atop many tons of high explosives. If it blew to bits, what would I do? I had been chasing cosmic wrinkles most of my career in cosmology. Would the chase end here, on the Pacific shore, under a fireball that rained metallic debris?

In the twilight, hundreds of us gathered outside the Delta site. Numerous buses were waiting. A thousand guests—one of the rocket base's largest turnouts ever—had gathered to see that rare spectacle: a dawn launch. While we watched, the countdown would be handled from the blockhouse, a protected structure about four miles from the launchpad.

The buses hauled us to the viewing area, a mile or so from the rocket. We got out and saw the Delta, brightly illuminated on its launchpad. Through the night, weather forecasters had launched small high-altitude balloons to check on upper-air winds; they were too fast for comfort, and there was a chance the launch might be canceled. Twenty minutes before launch, another balloon was lofted; its sensors indicated the winds had slowed down. That was promising.

I was standing close to Mike Hauser, a fellow principal investigator, and we exchanged silent glances, knowing that of the thousand people in the crowd, the two of us and John Mather had the most invested in the day. On an earlier trip I had seen the rocket up close, and had been aghast at how decrepit it looked, rusting here and there, patched here and there, spot repairs made with Glyptal. Our professional life's work was on top of that thing. We didn't say a word, only silent prayers.

The sky looked calm, clear, inviting. A ribbon of orange glowed in the east. Sunrise was moments away. My watch indicated 6:34 A.M. The launch window was the next twenty minutes, if the satellite was to enter the correct polar orbit, one parallel to the terminator, the shadow line between the day and night halves of Earth. The countdown came down to the last ten seconds.

"Ten, nine, eight, seven, six . . ."

Brows furrowed, muscles tensed; everyone held his breath. At such moments one fears the worst; would the rocket explode in a blinding flash?

". . . five, four, three, two, one . . ."

Someone joked, "If the rocket blows up, everyone hold George down or he'll kill himself!" I shot back: "Kill *myself*?"

". . . ignition."

The terrain lit up, as bright as the Sun. We could see the rocket lift off, in eerie silence, the launchpad engulfed in flames. Suddenly, seconds later, the sound hit with a roar, vibrating my chest. The air shook. The crowd gasped as the rocket continued its climb, spewing flame and smoke, growling all the way.

During takeoff, hundreds of thousands of gallons of water gushed underneath the rocket to suppress sound waves that otherwise might have reflected upward and damaged the rocket. Likewise, only six of the nine boosters were started until the rocket had risen well above the mountains; otherwise the reflected sound waves might destroy the craft. As we nervously watched the rocket's ascent, the second set of boosters ignited.

A half minute into launch, the Delta was traveling at the speed of sound—and still accelerating. We started breathing again. At one point it crossed the disk of the Moon. At 78 seconds, the rocket was more than 12 miles high and traveling at 2,744 feet per second. Suddenly, it seemed to expand, then expel six shiny specks—the solid fuel rockets. Their fuel exhausted, they fell back to Earth and landed 19 miles away, in the Pacific. At 124 seconds, the Delta was traveling at 6,207 feet per second and was 26.9 miles high; it ejected the last three solids, which hit the sea 231 miles from the launch site. At 240 seconds, the second stage ignited; the Delta split in half, dropping the first stage.

The fairings popped off at 245 seconds, 70 miles above Earth. Now COBE was racing through the cold and vacuum of outer space at 16,822 feet per second. The second-stage engine shut off at 650 seconds. The satellite coasted at 5 miles per second, 105 miles above the Earth's surface. COBE was in orbit. The third stage boosted COBE to its planned higher orbit.

"I must say, that was quite a ride," said a delighted Dennis McCarthy. At a postlaunch briefing, he said COBE was "very stable" in its orbit, 900.5 kilometers high by 899.3 kilometers low—almost perfectly circular.

I was elated. It had been a perfect launch. But we couldn't spend too long celebrating. It was time to get to the airport and go back to Goddard to monitor operations and run the experiment. Jet Propulsion Laboratory and COBE scientist Mike Janssen and I drove to his house, 120 miles away in Pasadena, to pick up his luggage on the way to the Los Angeles Airport. As soon as we arrived at Mike's house we called back to satellite operations and learned that COBE was right on target. John Wolfgang told us that the solar cells had generated extra power over the South Pole—that was the result of extra sunlight reflected from the ice. They were having trouble dissipating the power. Could they turn on the differential microwave radiometer early? "Yes," I immediately replied. This was a "Whatever you do, don't throw me in the brier patch" kind of question. (For months I had been arguing with the mission planners to turn the instrument on as soon as possible. They kept refusing, saying that we must first check out the satellite.) Up went the commands and down came the DMR signal. By great luck it immediately registered a spike, exactly the signal we had expected to see when the antenna crossed the Moon. Moments later our automatic on-board calibrator sent out a pulse. The DMR crew cheered. The first thirty seconds of data produced signals just like our ground tests and simulations. The DMR was working. It had survived launch. We gathered Mike's luggage, drove laughing and joking to the airport, and boarded a plane for Washington, D.C., smiling all the way to Goddard Space Flight Center.

Within hours COBE would begin telling us about the early life of the universe.

High above earth, with its three solar panel arms fully extended, COBE was nineteen feet long and weighed two and a half tons. It slowly rotated—reaching eight tenths of a rotation per minute by the end of the week—to allow each instrument to map the sky. The rotation also allowed even solar heating of the satellite. Most of the time, COBE stored its data on two on-board tape recorders, then transmitted the data once daily to the ground station at Wallops Island, Virginia. Then Wallops transferred the data to COBE mission control at Goddard. Often we listened in live, collecting real-time data through the TDRSS (Tracking and Data Relay Satellite System). Initially, we did this enthusiastically, excited that, after the years of setbacks and anguish, our instrument was actually doing the job. As the data began to build up, providing us with a slowly coalescing picture of the early cosmos, this periodic eavesdropping became less compulsive.

Arriving at Goddard in the early evening the day of the launch, I immediately went to the instrument operations room. There, teams for the three instruments (the DMR, FIRAS, and DIRBE) were sitting at a large Y-shaped set of desks, monitoring the satellite's operation and preparing to work with the flow of data. Pete Jackson and Charlie Backus had run off plots of the first half hour's data from the differential microwave radiometer. The plots showed the Moon signal, noise source calibration, and data. We all signed the first plot, made copies, and pasted them on the walls. And at the end of the first full day of data flow, we coaxed the analytical software to produce a rough but recognizable celestial map of the cosmic background radiation. Sergio Torres projected the map on the big screen, and the instrument operations room filled with applause.

The map didn't show anything we hadn't expected. There were no glimpses of the elusive wrinkles, the cosmic seeds we so urgently sought. At this stage the map was barely decipherable; it was like trying to see outlines of objects in a thick mist. It would take time for the mist to clear, we all knew that. The differential microwave radiometer was working as it should. The applause was as much an expression of relief as a recognition of our achievement.

The first few weeks were exhilarating, exhausting, and punctuated by a brief but very real scare. Three days after launch, we got indications that one of the gyroscopes, which helped stabilize the satellite, wasn't working. At about 2:30 A.M. John Mather

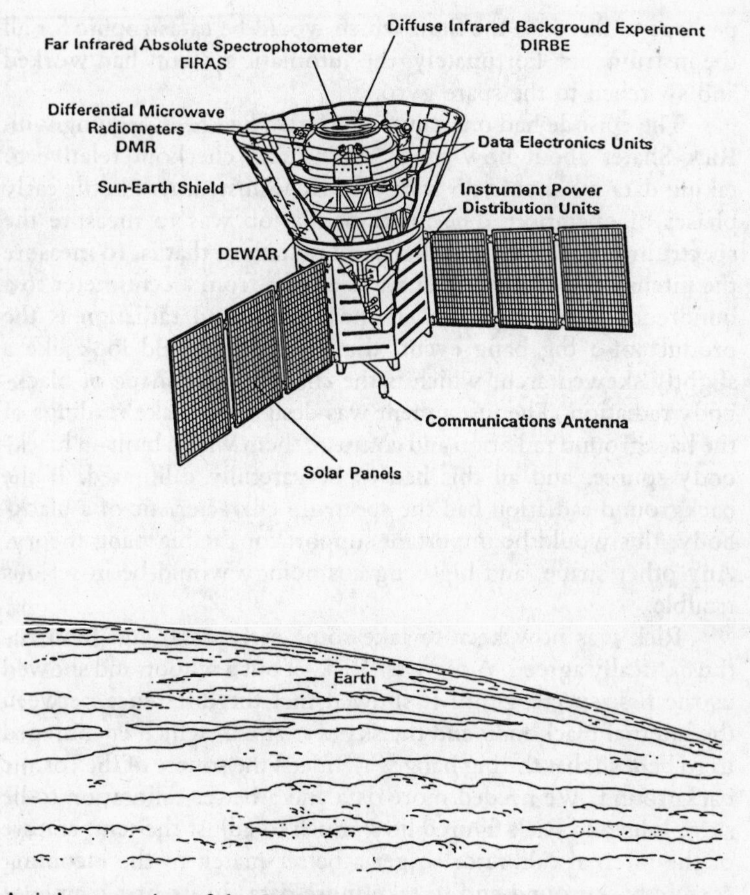

Artist's conception of the COBE satellite in orbit. The DIRBE, DMR, and FIRAS experiments are indicated. NASA/GSSC

and I had left to go home and rest; I had hardly gotten into bed when Rick Shafer called to tell me of the gyro failure and an emergency meeting. John Mather got the news through a call to his home, at four in the morning. Driving to Goddard in the pre-dawn darkness, he thought: If it's dead, it's dead. If not, we need to think carefully about what to do. He was worried that the loss of the gyro might destabilize COBE, causing it to tip over—

perhaps to face into the Sun, which would be catastrophic for all the instruments. Fortunately, the automatic systems had worked and switched to the spare gyro.

The episode had one salutary effect. I had been arguing with Rick Shafer about how much testing and checkout relative to taking data we should do with the FIRAS instrument in the early phases of operation. The instrument's job was to measure the spectrum of the cosmic background radiation, that is, to measure the intensity over a range of wavelengths from a centimeter to a hundredth of a centimeter. If the background radiation is the product of a big bang event, the spectrum should look like a slightly skewed arch, which is the characteristic shape of blackbody radiation. The instrument was designed to take readings of the background radiation and compare them with a built-in blackbody source, and all this had to be carefully calibrated. If the background radiation had the spectrum characteristic of a blackbody, this would be important support for the big bang theory. Any other shape, and big bang cosmology would be in serious trouble.

Rick was now keen to take some early data, and John enthusiastically agreed. A day later Rick let out a whoop and showed us the first spectrogram. It showed that the difference between the internal blackbody and the sky was small, which encouraged us to believe that the big bang was indeed the source of the cosmic background. We needed more data and a better calibration to be sure. John and Rick figured how much to adjust the temperature of the internal calibrator to get a better match to the incoming cosmic background and to take more data. In its first couple of weeks, FIRAS had collected nine minutes of observations, looking toward the north galactic pole, where there isn't much galactic dust. The instrument's accuracy was one part in a thousand, which would be a hundred times more accurate than anything before. Fragments of the spectrum had been collected over the years, beginning of course with Penzias and Wilson's original measurements, in 1964. Fragments, however, were not complete proof, and that is what those nine minutes of FIRAS data offered, for the first time.

We had made arrangements to talk about COBE at the American Astronomical Society meeting in Arlington, Virginia, in January 1990, just two months after launch. If we could get ready in time it would be the perfect place to announce the results we had to date. Beforehand, John recalls, "We were all totally exhausted. We made up our minds we wouldn't say anything until the publication was in the mail to the *Astrophysical Journal*."

The day of our talks arrived. John and I drove down to the conference hotel together and walked across the street to drop the manuscript in the mailbox, addressed to the editorial offices of the *Astrophysical Journal*. That was an important moment, as much symbolic as it was practical, consigning the first results of the COBE venture to the scientific press. It was the conference's last day, Saturday, and so we expected a low turnout. When we walked into the room we faced at least a thousand people. Palpable anticipation filled the room, an expectation of something important. Dave Wilkinson, a member of the COBE team, could barely contain his pleasure. Sworn to secrecy, as we all were, Dave hadn't been able to resist showing his Princeton colleague, Jim Peebles, a small scrap of paper. Jim had looked at the graph on the paper and shared Dave's pleasure.

John took the podium. After a brief introduction, he projected a theoretical graph of blackbody radiation, showing how the cosmic background radiation should look if it truly emanated from the big bang. Imposed over the theoretical curve were the sixty-seven data points, measured by FIRAS, strung together to yield the actual spectrum of the background radiation. The measured points matched the theoretical curve precisely, without deviation. Measurements in science, particularly in astrophysics, often have a large degree of "wobble" in them, so that a theoretical and actual curve may be close, but not perfectly matching. Yet, this matched perfectly—our errors were very small.

A moment's silence hung in the air as the projection illuminated the screen. Then the audience rose and burst into loud applause.* Such spontaneous displays of enthusiasm are rare in science, as most practitioners like to weigh what they are shown, and then make a cautious judgment. In this case, the data were so cogent that caution was unnecessary; a judgment could be made

*Recall that Slipher received a standing ovation even though no one understood what his data meant.

on the spot. But there was more to it than that, I believe. Here we were in 1990, forty-two years after the prediction of the cosmic background radiation, twenty-six years after Penzias and Wilson had first detected it, and yet there lingered doubts about the validity of the big bang theory. The Berkeley-Nagoya announcement in late 1987 of surprising spectral distortions in the cosmic background radiation, though questionable, had sown further seeds of doubt. No doubt about it, the cosmological community had become ever more edgy about the big bang. FIRAS had dispelled that doubt. Like the burst of applause that had greeted the posting of the first day's celestial map in the COBE instrument operations room barely a month earlier, the reaction to the FIRAS data reflected as much relief as recognition of good science. The big bang was still on track.

After John gave his report, I gave mine about the early results from the differential microwave radiometer. What I was able to show was extremely preliminary—the image had barely begun to coalesce, but it was still the best map anyone had. I was able to report that we saw the dipole, the result of the Milky Way's peculiar velocity with respect to the background radiation. This was important because it demonstrated that the instrument was working as it should. Apart from that, however, the background radiation looked smooth, a uniform signal from all points in the universe. We had no indication of wrinkles, no sign of cosmic seeds from which galaxies grew early in the history of the universe. This disappointment aside, our data clearly were consistent with the simple big bang model. More applause, though not the standing ovation the FIRAS results had received.

By a supreme irony, the session was chaired by Geoffrey Burbidge, of the University of California at San Diego, who was (and remains) a leading opponent of the big bang theory. As I finished my talk, I heard him audibly grumble that we and the audience had swallowed the big bang story—"book, verse, and chapter."

Mike Hauser then showed DIRBE's beautiful new infrared pictures of our galaxy, the Milky Way. It was poster material, dramatically showing part of the disklike galactic plane and the bulbous center.

Although the conference participants were enthusiastic about COBE's initial results, the news media focused more on our failure to detect anisotropy, the evidence of cosmic seeds. "Any structure in the early universe will leave an imprint that we should be able to see today," I said. But, I admitted, no imprint was apparent, so "we're still stuck with the question of where the galaxies come from. . . . It's really hard to understand why we don't see something like the Andes when we do the map. . . . If we don't see something, there's something wrong, really fundamentally wrong, with our theories."

John agreed, joking: "We haven't ruled out our own existence yet. But I'm completely mystified as to how the present-day structure exists without having left some signature on the background radiation."

Still, I cautioned reporters against premature pessimism. We would continue analyzing new data as they arrived. As information accumulated, the sensitivity would become more acute, I assured them, and then the ripples would turn up. Secretly, I had my fingers crossed.

Laurie Rokke, who joined our effort eight months before the COBE launch, was task leader for data analysis in the DMR data-processing team, based at Goddard. We ran into problems almost daily, both before and after launch, but managed to develop a software system that worked as desired. When we started out, it took four hours to process a day of DMR data, but we quickly got it down to thirty-eight minutes, depending on the computer we used.

In the early days after launch, we were there day and night. Laurie reminisced, "We would be in the DMR operations center, which was just a room in a trailer, for real-time passes. Then we would run back to our offices to finish coding software. Then we would run back to operations center for another real-time pass. We were doing this at all hours of the day and night." For the first few months we maintained around-the-clock coverage. Once operations were routine, the COBE satellite operations center took over the task of monitoring the real-time passes. We focused on longer-term trends.

I threw a party at my Maryland home after the launch. We kept computers, terminals, and modems there so we could monitor real-time passes. Early on, "we wouldn't miss a real-time pass for anything," Laurie recalls. "We would party, then run to the computer to monitor the real-time data. If it looked okay, then we would head back to the drinks, chips, and socializing."

Thirteen years had passed since that all-nighter at Berkeley, when Jon Aymon and I had seen evidence of the dipole on our computer screens, evidence from our U-2-borne DMR that our galaxy is being dragged through space by some distant, massive object. Daily, we saw the dipole again on our COBE-borne DMR. We were impatient to see new things—the seeds. But we knew they would be extraordinarily difficult to detect.

We were looking for tiny variations in the smooth background temperatures, something less than one part in a hundred thousand—that is something like trying to spot a dust mote lying on a vast, smooth surface like a skating rink. And, just like a skating rink, there would be many irregularities on the surface that had nothing to do with those we sought. These irregularities are like the systematic errors that plagued the differential microwave radiometer data—stray sources of heat, magnetic radiation, artifacts in software analysis, and so on. It is difficult to convey how obsessed we were with trying to eliminate these errors. I had started writing a list of potential things that could fool us back in 1974. Ever since I had continually updated the list, adding new candidates and studying their effects. I drilled everyone to check for errors all the time. And then I would have them double-checked by someone else.

When we'd been through that, I got Al Kogut in to check them through *again*. He would be someone fresh and independent of my years of effort. Al is an outstanding, capable scientist who is very thorough. That's why we trusted him. It was grueling work, hardly the stuff of glamorous discovery, but essential. We simply had to do it if we were to have any hope of boiling everything down to a purely cosmological signal, a signal that came from only one place—the edge of space and time. We had seen too many "discoveries" prove to be nothing more than artifacts and did not wish to be caught in the same trap. Al compared our venture to one of those British colonial explorers in the jungles of tropical India: He's pretty sure there's a tiger in that jungle, but he's not sure where. So he's very warily cracking sticks until he hears a roar. "Likewise, we were very warily cracking sticks until we heard a roar," joked Al.

About a year after launch, we heard the first barely audible roar. All of us knew that if we did discover cosmic wrinkles, they wouldn't leap out of the map overnight, fully formed. If present, their image on the map would slowly become more visible, for two reasons: First, the repetition of their exceedingly weak signal would become ever more evident, like the ever-darkening mark left by repeatedly rubbing a pencil lightly over a piece of paper; second, we would steadily clear out the noise in the system that tended to obscure the signal.

There would be four steps in the discovery of wrinkles: First, we would see the cosmic background radiation itself, seemingly uniform from all parts of the universe (as it appeared to Penzias and Wilson); second, we would observe the dipole, a slight distortion of the background radiation caused by the peculiar motion of our galaxy (which we identified via the U-2); third, we would detect the quadrupole (which Francesco Melchorri's and Dave Wilkinson's groups erroneously thought they had detected in 1981), which corresponds to the first cosmic distortion; finally, we would find the wrinkles themselves, the primordial seeds. At the end of 1990, a year after launch, we chalked up the first two steps, but had only set tighter limits on the last two. Now, early in 1991, I thought I could see step three emerging from the data—the picture was getting crisper.

In March 1991, I told the science team what we saw in the data, but cautioned that the noise and our limits on potential errors in the system were still as great as the signal. In other words, it could be wrong. I said the same in June at the next meeting. Ned Wright, who was working on the FIRAS data at the time, looked over the DMR maps and reported that they showed a signal resembling a quadrupole. The rest of the team were much more cautious; they knew how cranky instruments could be and how exciting data could fall flat the next day, so they wanted to continue analyzing more data before making any announcements. I agreed.

We kept absolute secrecy. We couldn't have anything leaking out about what we had found, until we were certain about what we had.

As we worked on, doggedly scrutinizing the data for systematic errors, marking putative anisotropies, tension began to build, both inside and outside the team. Inside, a conflict began to grow over how real our results were, and, more tricky to handle, how soon we should go public with them. Outside, our colleagues were wondering what was going on. Here we were, a year since launch, blessed with the most sensitive instrument for detecting cosmic seeds, and yet we had reported nothing—beyond the fact that, by January 1991, we had seen no evidence of seeds, and that any variations in the cosmic background radiation had to be less than several parts per hundred thousand. Had we failed?

Colleagues often asked me what we had found; I said we were still trying to figure it out. A report by the National Research Council was ominous: "If no variations are found at these increased sensitivities, then theoretical extragalactic astronomy will be thrown into crisis. Something will be seriously wrong—either with our theories of galaxy formation or with our understanding of the cosmic background radiation."

News media made hay out of the "crisis" that allegedly faced big bang theorists. While the articles tended to be fairly balanced, the headlines had a prematurely funereal tone: THE BIG BANG: DEAD OR ALIVE? asked a headline in *Sky and Telescope*. A student magazine distributed by *USA Today* ran a story titled GOODBYE TO BIG BANG THEORY? A *Science News* article was titled STATE OF THE UNIVERSE—IF NOT WITH A BIG BANG, THEN WHAT? Said *Astronomy* magazine: "Beyond the Big Bang—New observations cast doubt on conventional theories of how the universe formed." *Science* magazine tried to balance things with the headline DESPITE REPORTS OF ITS DEATH, THE BIG BANG IS SAFE, but acknowledged: "Even so, cosmologists are going to have to rethink a lot of what they thought they knew about what happened later."

Veteran foes of the big bang theory seized the opportunity to toss a few literary hand grenades. They were mostly longtime advocates of the steady state theory, such as Jayant Narlikar and

Hamilton Arp. Narlikar stressed in March 1991 that so far, "COBE . . . has found no evidence for any unevenness in the [cosmic background] radiation. These latest observations pose a serious problem for cosmologists dedicated to the big bang. . . . Avoiding confrontation with observations is scarcely the hallmark of a good theory."

As was to be expected, Geoffrey Burbidge expressed similar comments. Geoff's views couldn't be dismissed lightly, because he is a genuinely outstanding scientist, and I like and admire him personally, despite our cosmological differences. In an essay for the February 1992 issue of *Scientific American,* he argued that a key flaw in the big bang theory was the continued controversy over the value of the Hubble expansion rate. This is the value that expresses the rate of cosmic expansion—it is inversely proportional to the age of the cosmos. Experts have debated its value for years. It is important to resolve its exact value because an extremely young cosmos—say, less than 10 billion years—is hard to reconcile with the extreme age of some galaxies and globular clusters, thought to approach 14 billion years. Obviously the universe can't be younger than its constituents. And yet, he wrote, "The most favored version of the big bang model yields a universe that is between seven and 13 billion years old."

Big bang cosmologists "contort themselves" to explain the alleged discrepancies, and anti–big bang theorists have been suppressed by "a form of censorship," Geoff charged. He preferred a neo–steady state cosmology in which "continuous creation takes place in little big bangs. . . . In such a model the cosmic microwaves are generated by the galaxies and never coupled to them." Geoff also made a big deal about the failure to date to detect cosmic anisotropy: The "cosmic microwave background appears smooth to at least one part in 100,000, close to the level at which the big bang must be abandoned or significantly modified."

Unbeknownst to the outside world, by the fall of 1991 our celestial map was coalescing ever more clearly, with evidence of wrinkles becoming ever more persuasive, though not conclusively so. We had analyzed our data with new software and gathered a clean map. In September we held a major meeting on systematic errors—again—and resolved yet more potential traps.

I could tell that confidence was building, and that worried me. We have to be even more diligent, I insisted, and offered a free airline ticket to anywhere in the world to anyone who could prove the wrinkles were an artifact. I wanted to motivate the team to look for problems rather than assume everything was fine. If you get cocky, you get sloppy. In science, if you want to see an effect, it is all too easy to be seduced into believing you have. To quote Richard Feynman: "The first principle is that you must not fool yourself and you're the easiest person to fool."

Ned Wright, convinced we had a genuine signal, pressed ever harder for us to publish our results and let the cosmological community evaluate them however it wished. On October 9, 1991, he delivered a "preliminary paper" on his views at a Science Working Group meeting. Ned argued that we should offer an article on the data to *Astrophysical Journal*. He thought the team should publish the data so theorists wouldn't waste any more time wandering down blind alleys. Dave Wilkinson urged caution: He hadn't forgotten the quadrupole he thought he'd detected years earlier, only to see it vanish. Ned said the COBE data were more reliable than the balloon data, and that we should express our confidence by publishing it. COBE, he argued, saw the whole sky from the stable environment of orbit, whereas a balloon in the relatively unstable atmosphere saw just a small fraction of the heavens. He also thought it was important to "kick-start" the publication effort "because it can take a while to get things [published] in the *Astrophysical Journal*."

Ned and I had some pretty heated sessions over this, and more than once he stormed out of the room convinced, I'm sure, that I was being obstinate for no good reason.

In the end, the group turned Ned down; we wouldn't notify the *Astrophysical Journal*, not yet.* As a member of the Science Working Group, Ned was bound by its decision to keep things quiet for the time being. We would wait until a slightly revised version of the new software had reprocessed the first year's data,

*Ned, like Al Kogut, played a key role in convincing the Science Working Group that the results were correct and thus should be published. Ned developed several data analysis algorithms and made simulations that checked the DMR results independently.

and we could check through a few more potential errors. The new software had more checks and robustness built into it. We felt it was important to assess the results very carefully before taking the fateful step of announcing them. Even a tentative announcement posed all kinds of potential risks. There was a great deal at stake; we saw no virtue in announcing a result that would probably attract massive press attention, then might fall through and humiliate us all. It's difficult to appreciate how instruments can turn around and bite you if you don't watch them every step of the way.

Rumors began to course through the cosmological community, and more and more of our colleagues tried to pry out of us what we had found. At one point Dick Bond came up to me and said, "I hear you have a thirteen-microkelvin quadrupole in your data." I smiled and answered him evasively. Sometimes someone would bait me by asking, "I hear you found anisotropies at such and such a level," and I'd reply, "That's a little big, don't you think?" Likewise, at Goddard, Sasha Kashlinsky worked right down the hall from Chuck Bennett; Sasha frequently walked into Chuck's office and tried to bait him by saying something like, "So, Chuck, I hear you saw a big quadrupole," or "I hear you see something but *no* quadrupole." Chuck didn't bite the bait. From time to time I sent out electronic-mail messages to the team reminding everyone to keep mum about our results until we were more sure of them. My anxieties may have been justified: I don't know if the MIT physicist Alan Lightman had heard any rumors, but his 1991 book *Ancient Light* included a brief, cryptic reference to "very new observations" by an unidentified scientific team that "hint at positive detections of [CBR] unevenness at the level of one part in 100,000—which would provoke a great sigh of relief among theorists—but these observations have not been fully analyzed." They say that during World War II, even the great "secret" of the atom bomb project was the subject of rumors throughout U.S. academia. Who knew what people might be saying about the DMR as 1991 drew to a close?

In October 1991 we had the latest version of the map pasted on the wall in the instrument operations room. It looked unprepossessing, with red and blue blotches on a green background. I stood in front of it one evening, no one else around, and thought, Yes, this is it. This is what I've spent eighteen years looking for.

Although I had been among the conservatives in the debate about going public with our findings, I felt deep down that we had done everything right with our instrument. I thought the team had worked so well together, the analysis was so careful, and the search for systematic errors so thorough, that we *had* found the wrinkles, the Holy Grail of modern cosmology.

But the experimentalist in me argued for one more check, one more test that would eliminate a potential source of distortion: radio interference from our own galaxy. Over the years I had studied our galaxy many times for its possible interference in cosmic background measurements. Fifteen years earlier, working with my then graduate student Chris Witebsky, I had made models of the galactic signal while designing the DMR. The models were based upon radio maps and other surveys of the sky, plus some interpolation. We had updated the models over the years, keeping them as current as we could. Each year during our mountaintop measurements of the cosmic background spectrum, we had made scans of the galaxy and compared them with our model. Our data had generally agreed with the models. But we had never measured to this level nor checked the more sparsely measured southern sky maps.

I had carefully gone through the models and concluded that if the input maps were correct, the galactic interference could be only about 10 percent of the signal we were seeing. Chuck Bennett disagreed; he argued that galactic interference might be a significant problem. He and Gary Hinshaw began a systematic and independent check of our estimates. Chuck believed galactic interference might distort our map; I argued that the only way the map could be distorted by galactic interference would be if there was something wrong with the galactic signal maps. Earlier, Giovanni De Amici and I had made galactic emission measurements at the White Mountain Observatory in California, in conjunction with our long-wavelength cosmic background radiation measurements. The White Mountain results agreed with the models but couldn't convincingly rule out error in the galactic maps. And if there was any possibility of its producing an erroneous result, I knew I could not ignore it, no matter how I felt.

Once again, the southern sky was key. The galaxy is much more spectacular there and it is much less well measured, especially very far south.

The only way to tackle the galactic question was to make improved galactic observations in Antarctica, where my group already had established an observation base. The National Science Foundation had already approved my group's proposal for long-wavelength spectrum measurements complementary to FIRAS. The continent's cold, dry air, absence of pollution, low air pressure, and—away from the Transantarctic Mountains—huge expanses of flat terrain make Antarctica the next best thing to outer space, astronomically speaking. There, at an observation station set up just two kilometers from the South Pole, we would be able to collect electromagnetic whispers from the Milky Way that would confirm—or otherwise—the validity of our celestial map.

Chapter 13

An Awful Place to Do Science

Men wanted for hazardous journey. Small wages, bitter cold, long months of complete darkness, constant danger, safe return doubtful. Honor and recognition in case of success.

—Ernest Shackleton's newspaper advertisement calling for Antarctic explorers, circa 1900

To a first-time visitor, Antarctica can seem like a fantasyland: an ivory vastness bisected by ethereal mountains. Centuries ago Europeans speculated that a lush, inhabited continent existed at the bottom of the Earth, but explorers in the nineteenth century discovered differently. The continent froze them, starved them, drove them mad, swallowed them in great storms, and chewed up their ships with its icy teeth. "Great God!" wrote the explorer Robert Scott in 1912, shortly before the continent claimed him. "This is an awful place . . . !"

In November 1991, my colleagues and I ventured to this formidable realm. Reluctantly ventured, because I knew something of the physical conditions we would face. Extreme cold, high winds, high altitude, isolation, and the threat of serious injury or even death. I would add to Scott's desperate comment, and say of Antarctica, "This is an awful place to do science." Unfortunately, we had no choice in the matter: We wanted to chart the cacophony of radio emissions from our own galaxy, the Milky Way, and Antarctica was the best place on Earth to do that.

The cosmic background explorer satellite (COBE), launched exactly two years earlier, passed directly over Antarctica every hundred minutes at an altitude of some six hundred miles. Two full years of data flow from COBE, analyzed with laborious scrutiny back at the Goddard Space Flight Center and in Berkeley, had brought us to the brink of identifying wrinkles in the background radiation, the cosmic seeds from which the galaxies had begun to grow a billion years after the big bang. But there was just a chance—small, I thought, but big enough not to be ignored—that our map of the wrinkles was an artifact of the torrent of radiation that pours constantly from cosmic activity within the Milky Way.

Astronomers had charted the galactic radiation earlier, but too patchily for our comfort. We had to be completely certain of its profile so that we could compare it with our celestial map and take account of how the galactic input might distort the map. If we were going to announce the discovery of seeds from the dawn of time that, over the history of the universe, would have grown into present-day celestial structures, we had to be absolutely certain that we had made no mistakes. No, there was no avoiding Antarctica.

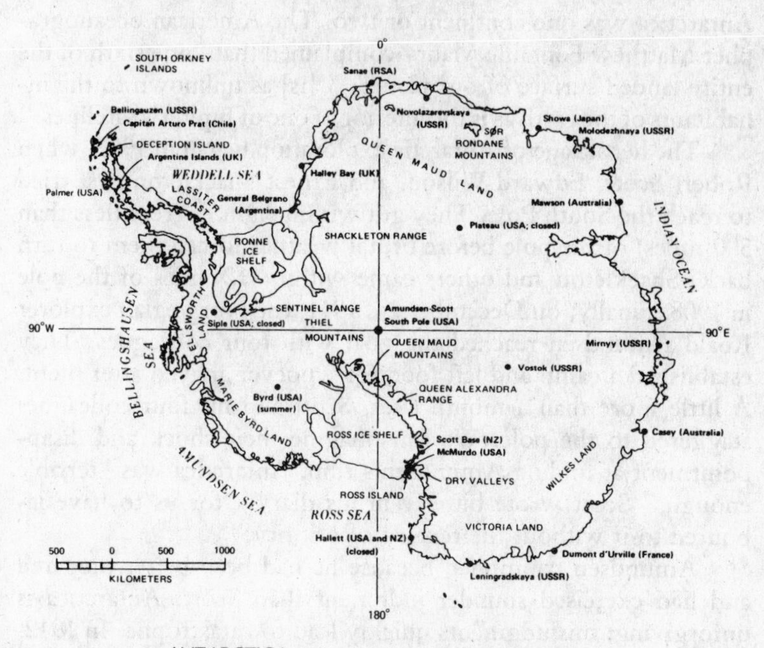

ANTARCTICA: selected stations and physical features.

* * *

Antarctica, which accounts for one sixth of the planet's land surface—size of the United States and Mexico combined—is a truly alien world: From its high, dry valleys to its vast ice shelves, the continent, once host to luxuriant life forms, is now sparsely populated, mostly at its fringes. The towering Transantarctic Mountains, which separate East Antarctica from West Antarctica, diminish and become buried under an ice cap more than a mile thick. The weight of the ice cap is so great that the continental rock beneath sags: remove the ice, and the continent would spring up more than 300 feet; melt the ice, and the global sea level would rise 180 feet.

Many historians believe the first person who positively identified the Antarctic continent was the Russian captain and explorer Fabian Gottlieb von Bellingshausen, on a sailing expedition in 1820. Victorian visionaries saw Antarctica as the last unconquered terrain on Earth, and campaigned for its conquest. They argued that science would benefit by answering questions such as whether Antarctica was one continent or two. The American oceanographer Matthew Fontaine Maury complained that "one sixth of the entire landed surface of our planet . . . [is] as unknown to the inhabitants of the earth as is the interior of one of Jupiter's satellites."

The heroic age of Antarctic exploration began in 1902, when Robert Scott, Edward Wilson, and Ernest Shackleton first tried to reach the South Pole. They got within eight degrees (less than 500 miles) of the pole before brutal weather forced them to turn back. Shackleton and others came within 112 miles of the pole in 1908. Finally, on December 14, 1911, the Norwegian explorer Roald Amundsen reached the goal with four colleagues. They established a camp and left food for whoever arrived after them. A little more than a month later, Scott and his four colleagues staggered to the pole. We can imagine their shock and disappointment at finding Amundsen's ruin. Antarctica was "terrible enough," Scott wrote bitterly in his diary, "for us to have laboured to it without the reward of priority."

Amundsen triumphed because he had been better prepared and had exercised sounder judgment than Scott. Antarctica is unforgiving; misjudgments quickly lead to catastrophe. In 1912, struggling to reach the pole, Scott unwisely relied more on ponies than on sled dogs. For this and other misjudgments, he and his colleagues paid with their lives. One of Scott's huts still stands on Ross Island, south of the Antarctic Peninsula. Inside you can smell his ponies' hay, preserved for eight decades like a corpse in an icebox.

Ironically, popular mythology remembers Scott more vividly than Amundsen. "The competent but prosaic Roald Amundsen seems to function chiefly as a foil for the tragic Robert Scott," observes the Antarctic scholar Steven J. Pyne. "Nowhere in Western literature is there a more compelling, sustained chronicle of life, humanity, and civilization reduced to their minima." I'm fascinated by history, and Scott's demise haunted me as we prepared for our expedition to Antarctica. Our venture to this hostile, frozen wasteland in November 1991 during the Antarctic summer, when the sun never sets, became more than a scientific expedition. It was also a personal test.

In Antarctica, a nice summer day can be minus 30 degrees; the windchill may mean instant frostbite. Growing up in Alaska didn't make me yearn nostalgically for numb earlobes and frostbitten fingertips. I'm not invigorated by freezing temperatures, and I don't gleefully don a Speedo and leap into an ice-speckled lake on New Year's Day. When the thermometer plummets my body begins to slow down; my muscles tend to crunch up; I fight colds and sinus infections, become feverish and tired, and feel just plain awful. Even worse, the South Pole is at high altitude and very dry. There is not enough air to breathe normally. Cuts get insufficient oxygen to heal. I expected Antarctica to be nonstop misery. When I confided my concerns to some of my friends, they replied, "You can stand *anything* for a month."

I couldn't forget the fate of Scott's expedition: the lifeless adventurers, frozen stiff with their eyes open, agonizingly close to their journey's end, within eleven miles of a cabin, warmth, and food. Had their legs held out a little longer, they would be a historical footnote today. Instead they died valiantly, needlessly, and so became heroes, mythic figures. Their expedition was the epitome of The Noble Failure. To a later, more cynical generation, they became something else—The *Perfect* Failure—as well as a handy literary metaphor (second only to the *Titanic*) for the twilight of Victorian optimism. Scott's dying words in his diary haunted me: "The failure of this expedition can in no way be attributed to the lack of planning and the efforts of the men." I had no interest in our becoming anyone's metaphor for disaster.

After our expedition was approved by the National Science Foundation, we attended safety orientation sessions. Antarctic officials showed us horrifying photos of what happens to people who forget to take proper care. For example, one man hauling a sled with a rope was so numb that he didn't realize the rope was cutting off blood flow to his hand. Result: His hand had a blue line across it, and everything from the blue line down to the fingers was black. His fingers had swollen so much the skin had split. They looked like overinflated balloons.

To be admitted to U.S. bases in Antarctica, we were each required to have full medical checkups. Officialdom didn't want any of its clients to require an emergency heart bypass or extraction of wisdom teeth while stuck at the pole. There were gruesome tales about people who got sick while stranded at other bases in Antarctica—like a doctor at a Russian station who was forced to cut out his own appendix. All U.S. personnel are required to have full mouth X rays so that, if worse comes to worst, their remains can be identified.

I tried to allay my fear by throwing myself into preparations for the trip. I would lead the expedition. Marc Bensadoun, a graduate student, and Giovanni De Amici organized the logistical operations. Giovanni was also in charge of measuring galactic emission. Giovanni, a tremendously energetic, mountaineering Italian who doesn't like wine or garlic, has a down-to-business air about him. The leader of our Italian collaborators from Milano, Giorgio Sironi, could not go as his preparatory medical exam had uncovered a potential tumor. He had an operation but was still recovering. Our Antarctica team had worked together previously, at White Mountain Observatory in California and in Italy.

For the six months prior to November 1991 the inhabitants of the U.S. base at the South Pole had lived in total darkness, and for eight months had been almost completely isolated from the world. That isolation had been broken just once, at midwinter, when a navy plane flew over to drop mail and extra supplies.

Now the long winter was ending and the Sun was rising higher and higher in the sky with every passing day. Four days before my departure, John Lynch of the National Science Foundation's Antarctic Program cheerfully informed me that the South Pole was officially "open" for the season. The surface winds were twenty-five knots and the surface temperature was a balmy 73 degrees below zero. Oh God: *−73°F and twenty-five knots!* I had never endured anything worse than −30°F. My fears returned with a vengeance. The day before departure in a panic shopping spree, I spent three hundred dollars more on additional thermal underwear. This included another heavyweight set and an expedition-weight set and then two lightweight sets to wear underneath in case it was really bad. I also bought a better face mask and "overmittens" for protection against the wind.

U.S. research teams reach Antarctica via a base in Christchurch, New Zealand. There, officials briefed us for our expedition and taught us what to do should the plane ditch in Antarctic waters. They weren't kidding: Unpredictable weather and the lack of air traffic control facilities make these flights risky. Across the ivory continent, tail fins of planes downed decades ago protrude from the ice like tombstones. Thrill seekers pay thousands of dollars for glamorous commercial flights over the pole that endanger their lives.

In our red parkas and big boots, we walked out onto the Christchurch runway toward the navy air transport, a C-141, which resembles a metallic whale with wings. Carry-on baggage was whatever you could stuff in a single duffel bag. There we were, in the middle of summer in New Zealand, and each of us was wearing our full survival gear, carrying a fifty-pound tote bag, and waiting in line on the tarmac. We sweated profusely. We each were handed earplugs and a box lunch. Inside, the aircraft was like a 747, but gutted—no comfortable seats, just the skeletal framework of the plane. It was spartan, but it eliminated a lot of needless weight. We strapped ourselves into red webbing strung along the wall. Looming over us were giant wooden boxes of cargo, mainly supplies and heavy construction equipment such as backhoes, plus our scientific gear. The cargo took up so much room that we had to stuff ourselves into whatever space was left.

There wasn't enough room for your feet; you had to be friends with your neighbors so they would let you stretch your legs under their "seat."

The engines revved, we bumped down the runway, and we felt that sudden, brief weightless feeling. We were in the air, bound for McMurdo Station, Antarctica, over thousands of miles of choppy, chill waters. First-time visitors to Antarctica were visibly nervous, whereas old hands were acting like . . . well, like old hands. The plane was so noisy that you couldn't talk to your neighbor without shouting, so in went the earplugs. This was a ritual for all Antarctic flights. For the next eight hours, we could read, sleep, look out of the window, or eat our snacks—sandwiches, fruit juice, and desert. That's it. Most of the time it was just plain boring. I was tired, so mostly I slept. Occasionally I glanced out the window and wondered when we'd see icebergs.

An hour before we landed at McMurdo, I felt both excitement and trepidation. I wasn't the only one. A woman with the air force started putting on more clothes. Within fifteen minutes of landing, she had donned three hats plus liners, inner gloves, and overmittens. The other passengers watched her out of the corners of their eyes and nudged each other.*

Landing on ice feels like landing on concrete: Bump, screech, and you're down. The only difference is that the passengers glance excitedly at each other and move their lips as if to say, "Holy cow, we're in Antarctica." We had landed on a frozen stretch of the Ross Sea. Through the window came a brilliant white glow. The side door popped open; cold air blasted in. Lugging our fifty-pound sacks, we edged our way down the ladder, down onto

*I flew down with Steve Levin, a former graduate student on the project, who now works at JPL on the Search for Extraterrestrial Intelligence (SETI). We got through to Antarctica right away. Part of the team—Giovanni, Marc Bensadoun, Michele Limon, and Marco Bersanelli—had a more typical experience. Their flight was canceled repeatedly, once stranding them in the warm Christchurch weather wearing arctic outfits. After many very-early-morning disappointments, they flew south—only to have to turn back halfway there. Marco was asleep when they turned back and woke shortly before landing. He quickly dressed to the maximum in his Antarctic survival suit and looked out the window. "I didn't know Antarctica would be so green," he said before noticing no one else had suited up. Antarctica is definitely not green.

the crunchy snow. Around us was an alien landscape. Whiteness almost everywhere, glaciers so immense they disappeared into the clouds. We felt breathless with exhilaration; it was as close to a transcendental experience as one can have on Earth. The Sun was so bright it was disorienting; I felt fuzzy and askew as I fumbled to put on my sunglasses. In the distance I could see the Transantarctic range. I started taking photographs. My hands were shaking in the cold wind. The C-141's motors were kept running, lest they freeze, which made it more windy.

Our awe faded as we realized our skin was growing numb. It was about 20 degrees below zero Fahrenheit. In the distance, men in uniform were yelling at us to get out of the way of the plane and to board their transport vehicle, which resembled a great orange box on giant tractor tires—a trailer to a big-rig cab. Still gazing at the fantastic scene around us, we stumbled over to the vehicle, known as a Delta, and clambered up the slippery, steep steel stairs. As we departed, the rear of the C-141 yawned open like a giant clam. Forklifts rolled aboard, grabbed boxes, and hauled them away to a storage yard.

Traveling about ten miles per hour, the Delta headed toward our stopover—the town of McMurdo Station. McMurdo, the largest of three permanent sites the United States maintains in Antarctica, was established in the mid-1950's. With a population that fluctuates between 250 in the winter season to 1,200 in the summer, McMurdo is a scatter of huts and buildings of all shapes and sizes, and until recently had the reputation for being the most polluted spot on Earth. Decades of refuse of all kinds had piled up on the ice. In response to protests by organizations such as Greenpeace, an effort has been made to clean up the mess. It is still not pretty, and one National Science Foundation official once compared it to a mining camp. There are dirt roads, tractors, and strange vehicles rumbling around, and numerous barrels and crates scattered across equipment yards. Thank goodness we would be there for only three days, preparing for our trip to the South Pole Station, where we would set up our equipment, another thousand miles toward the center of the continent.

Wildlife exists almost solely on the Antarctic coastline. The McMurdo base is just far enough inland (connected to the sea by an inlet that's usually frozen over) that it appears to be lifeless except for humans. In reality, it brims with life—but you have to make an effort to see it. While there, we visited this hidden world. Halfway between McMurdo and Scott Station, there was an exploratory hole drilled into the ice, with a corrugated pipe lining and viewports at the bottom. To reach the hole, we took a half-hour hike out onto the ice. It was next to a natural hole in the ice where giant seals came up to sun and get warm. I removed my parka to descend the tube, which was uncomfortably narrow; there wasn't enough room to bend my knees in order to step down the rungs. Instead, I had to lower myself by my arms from one rung to the next. In that clumsy way, I descended the full forty feet. At the bottom, I looked through a porthole at the frigid, crepuscular, aquatic world beneath the ice. In the eerie bluish light that leaked through the ice, I could see plants on the ocean floor swaying gently in the current; jellyfish and shrimplike krill floated by. This was to be our last glimpse of life in its natural environment until we returned from South Pole Station.

A C-130 ski-plane took us from McMurdo to the South Pole, a flight much more spectacular and dangerous than the flight from New Zealand to McMurdo. En route to the pole, you fly over the rugged Transantarctic range, where glaciers thousands of feet thick ripple with deadly crevasses. Most of the time you're flying over a pure white expanse of absolutely nothing—as flat and meaningless as the surface of a billiard ball. First-time visitors to the South Pole—the bottom of the Earth—often get so excited that on arrival they jump up to run outside, which isn't a good idea. The pole is nine thousand feet above sea level, so the air is thin; just standing up and moving around quickly can make you dizzy. Overexert yourself and you might faint from altitude sickness, or you might suck in too much cold air and sear your throat.

Amidst the racket of grinding airplane engines, I could see the ceremonial pole a couple hundred feet away. It resembles a red-and-white barber pole topped with a silver sphere the size of a basketball, adjacent to a semicircle of flags blowing in the wind; visitors pose for photos that later decorate their living room walls. It is a wonderful visual metaphor. The ceremonial pole isn't far from the true, or geographical, pole, but the geographical pole is harder to mark because the ice keeps shifting position. While

George Smoot at the ceremonial South Pole. Note the frozen breath icing his beard. GIOVANNI DE AMICI

there, I took a video of a walk from the base to the ceremonial pole. If I ever need reminding of conditions in Antarctica, I simply have to watch the video and listen to my labored breathing and breathless attempts to describe the scene as I crunch through the energy-sapping snow. Usually, I don't need reminding.

A quarter mile away is South Pole Station, a U.S. base for scientific studies. At its center is a large, silvery geodesic dome streaked with snowdrifts. Inside the dome are smaller buildings, including a cafeteria, infirmary, sauna, library, video room, pool hall, and mini-gym. The dome is partially buried in the snow, so you have to walk down a ramp of crunchy ice to get inside. As you disappear into this tunnel and connecting caverns, you escape the wind and immediately feel warmer. Yet it is still a few tens of degrees below zero inside. A weird blue light glows through the few clear panels of the dome. Ropes of icicles hang down. The icicles crystallize from water vapor rising from people's breath. After an orientation meeting, pole officials assigned us bunks in a summer camp consisting of dark green tents that house many of the hundred-odd residents during the polar summer.

One of the first things you learn is: Don't put your boots on the floor overnight, otherwise they will be colder than ice by morning. You also learn the "three cases of beer" rule: If you stack three cases of beer in the tent, the bottom case will freeze, the next will become too cold to drink, and the top one will be just right.

Now it was time for science.

My spirits fell when we reached the observing site. It was nowhere near ready for use. Bad weather and sickness had prevented the station crew from getting it ready. The research tent was full of holes and partly buried under snowdrifts. So we had to clean up, prepare the site, and get our equipment hauled out there, unpacked, and set up. It would take days just to figure out where all our equipment had been stored; some was at the site, some was in the cargo yard, and some was still back at McMurdo or God knew where else.

Our first day working at the site was sheer, unadulterated misery. The cold gnawed at our skin, and the wind lashed it. We

Our 1991 Italian-American team—Michele Limon, Marco Bersanelli, Andrea Passerini, Giuseppe Bonelli, Giovanni De Amici *(kneeling)*, Bill Vinje, Marc Bensadoun, George Smoot, John Gibson *(kneeling)*, Steve Levin—at the geographic South Pole. GIOVANNI DE AMICI

tried to avoid touching metal outdoors with our bare hands, lest we suffer skin burns. (I couldn't forget that photo of frostbitten, blackened hands.) We also had to avoid overexerting ourselves. The pole is such an awful place that it quickly disorients and exhausts you, and disoriented, exhausted people make mistakes, sometimes fatal ones. Fortunately, because of our experiences at the White Mountain Observatory, which is at 14,000 feet, most of us were used to working at high altitude. We knew how important it was to walk slowly and to drink a lot of water. Occasionally you forget—but Mother Nature quickly reminds you. I got one such reminder. Late on the first day, when we were moving crates of equipment, I worked too hard for too long. The cold had made my nose runny, and as I sucked in air it froze the liquid in my nose. Suddenly I realized what was happening and began gasping for air. The gasping made things worse: The sudden rush of cold air into my lungs seared my throat. I stood there, wheezing and coughing. When I regained the feeling in my throat, it was sore and burning. I bitterly thought about my friends' reassuring comment: "You can stand *anything* for a month."

Already my worst fears seemed to be confirmed: We were way behind schedule. Critical pieces of our radio telescope dish and supplies were missing. And to top it all off, I was a wreck physically, a forty-six-year-old man having trouble breathing and eating and trapped at the bottom of the world. I fretted: Could things get worse? Should I just die now and save going through the rest of the pain?

Finally, the missing components of the dish arrived and we could start setting up. Giovanni and I took on the task of assembling the instrument, with which we would listen to the outpourings of the Milky Way. The dish had been built back in Berkeley in a rush. We had thrown everyone into it: We even brought our master instrument builder, Hal Dougherty, out of retirement; we'd hired undergraduate Christian Carter. Hal, Christian, John Yamada, Doug Heine, and other members of the group had worked around the clock. The main part of the dish consisted of twenty-four separate "petals" and twenty-four matching extension panels, which we called Christian's halo. When assembled, it would form a one-ton dish more than thirty feet wide. It had been a rush job, and although we knew it operated in the sunny climes of northern California, we were uneasy about how it would perform in Antarctica, which is as cruel to machines as it is to people.

Outdoors, our eyes streaming with tears from the cold wind, Giovanni and I assembled the dish piece by piece, petal by petal. Each of the twenty-four metal petals was connected to a central hub. It was like piecing together a sunflower. It was held together with hundreds of small screws and nuts, which we struggled to insert with our bare, numb hands. We knew the finished structure would be difficult enough to manhandle under the best of circumstances. Under Antarctic conditions it would be worse. As we worked, periodically we had to go inside to have a hot drink or thaw out our masks and gloves. Our breath froze on our face masks until they became stiff as a board. My goggles iced over until I couldn't see; I had to go inside repeatedly to clean them off. We warmed our hands over the melting ice on the stove.

Our labors wouldn't end when the dish was assembled. We would have to point it at different parts of the sky. In this regard, I had conceived of a hydraulic system that would tilt our radio telescope dish to different angles above the horizon. Unfortunately, we found out immediately that hydraulic fluid froze at the South Pole. So Giovanni and I improvised: We came up with a system using a bulldozer jack and sets of wooden blocks, which could be inserted or removed by hand to hold the dish at desired angles. We were concerned about the effect of the wind on the dish, so we got a bulldozer operator to carve a pit, shaped like a clamshell, in the ice. We planned to assemble the dish within the pit, which would act as a windbreak. We also set up fences around the site to lessen the effect of the wind. The pit was sloped so that it was deepest on the side from which the wind was blowing. But not too deep—we had come to the South Pole partly for its flat terrain and didn't want the dish to be confused by signals from the horizon.

We decided to secure the dish by freezing its supports and hinges into the ground. To do this, we first buried wood in the ice. Next, we melted ice in a five-gallon bucket on the oil stove. Then we took it outside and poured the water on the ice to make "concrete pads"—a solid base for the dish. We then moved the hub on its pallet outside and into the pit and bolted the hinges to the wooden blocks set into the ice concrete.

Finally we were ready to test the tilt system. But we ran into a serious problem. When Hal had made the hinges out of big inch-thick plates of aluminum back in Berkeley, he had not foreseen that we would have to move them on the shipping pallet. One corner of a hinge stuck up into the dish struts and was blocked from opening. At this point, tilting the dish seemed like a huge insoluble problem—my original hydraulic scheme was off and the new scheme was blocked. Giovanni wanted to disassemble the damn thing and drag it back indoors for alterations. No way, I said. If we did that, we wouldn't finish in time to get data. We could not take any more delays and hassles. We had only so much time in Antarctica; we had to make that dish work in the time left, because I'd be damned if I was going to go through all this and not get anything. The mere thought of ever having to come back to this godforsaken place . . . So I grabbed a hacksaw and stormed outside.

The wind was howling and the temperature was 40 degrees below zero. I was determined to solve the problem by sawing four to five inches off the hinge so that it would clear the struts and we would be back in business. I felt like the lunar astronaut who finally tried to fix a cranky gizmo the good old-fashioned, all-American way: He banged it with a wrench.

So there I was, sawing, panting, and muttering to myself about the damned radio telescope and Antarctica. I didn't realize I was breathing too hard. I was swallowing big gulps of polar air. The chill air spiraled into my lungs, searing my throat on the way down. It was so cold that my fingers and head grew numb, while the rest of my body—all bundled up—began to sweat. It was a perfect recipe for a medical collapse. And collapse I did. For the next couple of days, I was stricken with a fever and was of little use, except for washing dishes and other "indoor" work.

Finally, after two weeks of physical suffering, ill-temper, and hit-it-until-it-works, we were ready to eavesdrop on the Milky Way, a typical, medium-sized spiral galaxy containing 100 billion or so stars. Earth orbits a star in a distant corner of the galaxy, whose disk arches across the night sky like a cloud of fireflies, and whose center—somewhat obscured by dust clouds—glowers in the direction of the constellation Sagittarius, the Archer. From that disk, and particularly from the center, comes a clamor of radio noise generated by heat produced as dust clouds condense into planetary systems, as nebulae collapse into fusion-energy machines called stars, as neutron stars spin madly and suck matter from companion stars, and as cataclysmic events of an unknown nature (a monstrous black hole? antimatter annihilation?) transpire within the galactic core, where the stars crowd so tightly that night never falls.

There is another dramatic source of galactic noise: synchrotron radiation, emitted by high-energy electrons "complaining" as they are bent by stray galactic magnetic fields. Galactic synchrotron radiation was what worried us. It was so intense that COBE's search for cosmic ripples was like trying to hear a whispered conversation at a wild party. We had estimated galactic noise could cause about 10 percent of the signal in COBE differential microwave radiometer data. Ten percent may not sound like much of a problem. But we feared it might become a problem because the ground-based maps of synchrotron emission were patched together from several different radio telescope surveys. Since they didn't all match together, adjustments were necessary in order to merge them. And then we had to extrapolate the resulting merged map to our DMR frequencies; in so doing, significant errors might creep into the data. The true error might be much larger than 10 percent—perhaps large enough to account for what we thought were wrinkles. Once we had tracked down, scanned, and cataloged every source of galactic interference, we could "subtract" it from COBE's data and, we hoped, spot the long-sought cosmic seeds.

The contrast between the lofty scientific goals we sought and the means by which we sought them was great—not to say hilarious. Astronomical observations of this importance ought to be gathered on high-tech machines, driven by whirring motors, with a computer orchestrating automatic scanning of the cosmos. And here we were, about to scan the sky by moving the giant dish using ropes, wooden blocks, a bulldozer jack, and a lot of physical exertion by two crazy men, sometimes laughing together, often cursing each other. Welcome to science in Antarctica.

The first time we tried our system was a thrill. I loosened the ropes and Giovanni started cranking the jack. The dish began to tilt to its scanning angle. As soon as it got to 15 degrees, I slipped the 15-degree blocks in place. Then we tied the dish into position and ran inside the instrument tent to check the computer screens, which showed the signal rising and falling as the Earth rotated, causing it to scan different parts of the radio "sky." It worked! We hugged each other in triumph.

After some hours of taking data at that scanning angle, we wanted to see if we could scan at even steeper angles. So we went outside again and manned the ropes and cranks. We tilted the dish to 20 degrees. No problem. More data collection. Then on to 30 degrees, where it would scan across a large, intense radio source on the galactic plane. I was ecstatic. We're doing radio astronomy from the South Pole! I thought happily as I loosened the ropes. Giovanni was as chipper as I was, and he started cranking the jack. Suddenly the dish lurched. I was holding two of the ropes, and the dish swung forward, dragging me toward the pit. "What's going

Giovanni De Amici beside the assembled thirty-foot-diameter dish. GEORGE SMOOT

on?" Giovanni yelled. The one-ton dish fell on its side with a sickening crunch. I was shocked, and scrambled around to see if Giovanni had been crushed under the huge weight of the dish. He was nowhere to be seen. Fortunately, he had fallen into the pit next to the dish. "What happened?" he gasped as he emerged, breathless in the high altitude. "Did the wind blow it over?"

I felt dizzy. "No," I said. "I was just holding on to the ropes and letting them out lightly in case of a gust, when it started pulling hard." We had been careless, the thin air eroding our judgment, and someone could have been badly injured—or worse—as a result. It was a salutary reminder of the dangers we faced.

We had tilted the dish past its center of equilibrium, so gravity had set in and the dish fell over. Giovanni and I clambered around the dish, examining the petals and struts, but found no serious damage. We were lucky. Given how difficult and time-consuming it is to get spare parts at the South Pole, it would have been humiliating if we had brought our experiment to a premature end by ignoring the simple laws of physics.

We began routine collection of galactic data, and I was feeling pretty upbeat. I had confronted my worst fears and survived, and the equipment was working. It was a hopeful time.

Then one morning I was sitting in the instrument tent, watching the data flicker across the screen. "Oh my God," I blurted out, "it's aliens." Everybody in the tent turned around and looked at me as if it might be true. Even I half believed it. There on the screen, in the midst of a fairly even signal, came a huge spike. It curved up and down the screen, which is exactly what you'd expect for a signal from a point source passing overhead—a point source like, say, a flying saucer. "What on earth is that?" I asked rhetorically.

"Beats me," said Giovanni.

"Beats me too," I said.

Several times a day, the signal would show up on the computer screen, last a short time, then disappear. The signal clearly wasn't natural, like an aurora, for example; it was too spectrally pure, with a narrow-frequency signal, to be an aurora. I don't believe in UFOs, but after all I'd been through in the previous few weeks, I was thinking there was a conspiracy to prevent us from getting the data. I joked to Steve: "Here we took you from your SETI [Search for Extraterrestrial Intelligence] project to check the cosmic and galactic background, and now the damned aliens won't let us get the data." We considered a number of alternative ways to explain the errant signals, all far less dramatic than ETs. Perhaps our equipment was faulty and experiencing periodic gain fluctuations. Perhaps our dish was detecting stray signals from another team's experiment at the South Pole. Perhaps we were detecting fluctuations in the use of the local electric generator. The latter possibility especially interested us because the Princeton group led by Jeff Peterson was sharing a generator with us, and from time to time they switched on a huge vacuum pump.

After a few days we determined the signals tended to recur in periods of one hundred and something minutes, which suggested they might be from a satellite as it regularly passed overhead. But there was more than one type of signal, which puzzled me: If it was a satellite, then why didn't it have a consistent hundred-minute period? I pointed our dish in different directions, trying to identify the source. No luck. For a while the strange signals drove us crazy. I thought the interference would wreck all our observations. The repeated signals were simply trashing much of our data.

Within two or three days we hit on the likely culprit: not one satellite, but three. Although we never proved this explanation beyond a reasonable doubt, it strikes me as the best solution. One piece of evidence that there was more than one satellite was that the aliens didn't always show up at the same time; sometimes they signaled us at the same time of day and sometimes they didn't. If they really were separate satellites in separate orbits, that would explain the apparent unpredictability of the signals. Also, they were broadcasting in a band forbidden by international agreement.

The damage was done: Our dish had five frequency bands, but thanks to the aliens we obtained good data for only two of them. Hobbled like this, we continued collecting data for two more weeks, making a total of three in all. We were able to make a preliminary check of our galactic map against the existing, patchy maps, and as far as we could tell, they matched well. A more detailed comparison would have to await our return to Berkeley.

The galactic mapping work that Giovanni and I spent most of our time doing was a crucial check on the validity of our COBE data—what we thought were cosmic wrinkles. Mapping the galaxy is a good project in its own right, so we had no trouble maintaining secrecy about our immediate intent. We still didn't want news about our possible discovery of wrinkles to leak out prematurely. As I was in constant contact with Berkeley and Goddard via electronic mail, this proved tricky.

At the South Pole, E-mail is public, not private. Anyone could have read my messages to the United States or vice versa, and an astronomically savvy researcher—say, Jeff Peterson of Princeton—could have instantly put two and two together, had he stumbled over an explicit message between me and members of the team. So I had to choose my words gingerly. In Greenbelt, Al Kogut exercised similar caution in communicating with me. He sent me status reports saying things like, "We've got this and that program running, Phil Keegstra has run models X, Y, and Z, and the magnetic effects are not important"—very dry stuff, and certainly nothing like, "Hey, the anisotropy is plus or minus 2 today!"

Toward the middle of December, a ham radio operator connected me with my parents back in the States. I had forgotten about the time difference and awakened them at two in the morning. It took them a while to realize that the crackly voice on the other end of the telephone came from Antarctica, and that the voice was their son's. Mom asked: "Is it cold down there?" I assured her it was. Dad joked: "Don't you have enough sense to stay away from there? I took you to Alaska so you could see that's what that kind of place is like!" I told them that our work was going well, which was fortunate, as soon I would have to go back for a COBE review. Dad later told Mom he sensed something in my voice. "He's excited about something," he told her. "He hasn't told us everything."

Working at the South Pole has some transcendental qualities to it—especially when you are outside and alone. Out where we were, two kilometers from the pole, it is very flat, bright, and isolated. You have to look hard to find the station in the distance. It can be like a surreal dream, you and your instrument alone in a vast whiteness. You have no outside distractions. You can focus completely on the moment. You must pay attention to what you are doing. The experience was like fasting on a religious retreat or being in a sensory deprivation tank, but with a keen edge. With several different experiments under way, each member of the team had this experience. Michele Limon, suffering from frostbite to the end of his nose, came stamping into the tent slapping his hands together to shake off the cold and warm up. Asked how he was, Michele replied, "I feel so alive here. I love it in spite of the misery." He was right; there was something pure and exhilarating about the experience.

In mid-December, Steve Levin and I packed and prepared to return to the United States. I had to be back for what I hoped would be a final meeting regarding systematic errors in the COBE differential microwave radiometer data, and, of course, to do a more detailed analysis of the galactic mapping data we had been collecting the past month. At 4:30 A.M., December 15, the plane soared from the polar ice. I was buoyant on departure. We got to McMurdo and were scheduled to depart for New Zealand on a C-141 air transport that afternoon. At McMurdo, Steve took a welcome shower and nap, while I visited the National Science Foundation representative and filled out the appropriate paperwork regarding the expedition. There are forms everywhere. As we lifted off from McMurdo on the C-141, we could see that the ice shelf was starting to break up. That made the residents of McMurdo happy, because it meant a ship would soon be arriving to resupply the town. By that time, I would be back in the States, working on other things, and happily so. Steve Levin made me smile when he said:

The best day in Antarctica is the day you arrive.
The next best day is the day you leave.

Chapter 14

Toward the Ultimate Question

I stepped up to the podium shortly after 8:00 A.M., to give the first talk of the day, Thursday, April 23, 1992. The occasion was the annual meeting of the American Physical Society, at the Ramada-Renaissance Hotel, downtown Washington, D.C. Dressed in a crisp suit, I faced a large, expectant audience. We had let it be known that the COBE team would be making an important announcement, which fed a hubbub of anticipation as people settled into their seats. Only a very few people knew what we were about to say, though some surely guessed. Ours had been a well-kept secret. Soon, we would be able to talk publicly for the first time about our results, the hard-won fruits of an eighteen-year search. Experiencing a mix of relief and tension, I shuffled my notes.

In the few moments of suspense that hung between those routine, nerve-soothing preparations and the opening of my talk, I reflected on the fact that, almost precisely fifteen years earlier, I had made the first important announcement in my career as a cosmologist: the detection of the dipole in the cosmic background radiation. That detection, made by flying a differential microwave radiometer (DMR) aboard a U-2 spy plane, revealed that our galaxy is moving with a peculiar velocity of six hundred kilometers a second, under the gravitational attraction of some distant, massive concentration of galaxies. The universe, it turned out, is much more structured than anyone had guessed, with galaxies existing as components of large conglomerations rather than being distributed uniformly through space. Our U-2 experiments had also revealed that the cosmic background radiation is remarkably smooth, an apparently homogeneous afterglow of the big bang.

The fifteen years that had passed since the U-2 experiments served to reinforce these dual conclusions—thereby presenting cosmology with a conundrum. If the massive structures in today's universe formed under gravitational collapse during the 15 billion years since the big bang, then evidence of primordial structure *must* be visible in the cosmic background radiation, which gives us a view back to the universe at three hundred thousand years. Our spy plane–borne instrument was sensitive to differences in the background radiation of better than one part in a thousand, and yet it had detected no sign of incipient structure. By early 1992 the continued inability to detect wrinkles in the background radiation had become a serious embarrassment for cosmology and, specifically, for big bang theory. The DMR aboard the cosmic background explorer (COBE) satellite had boosted our detection sensitivity more than tenfold, and brought us close to the limit of what was technically possible. Had we failed to detect wrinkles with this instrument, the science would have been in deep trouble.

I had returned from Antarctica just four months prior to the American Physical Society gathering, pleased that I had met the challenge of the severe physical conditions of the pole. The galactic emission data that Giovanni De Amici and I had managed to collect supported the validity of galactic maps we had been using. We could continue our work with confidence.

Ahead lay the final obstacles to our mission's goal—some of them were technical, some psychological. We felt we had detected the long-elusive wrinkles, but I was keenly aware that we might still be the victims of a quirk of nature or instrumentation. Our map of the cosmic background radiation revealed fluctuations, with some regions warmer and others cooler than average. But we still could not be certain whether the pattern was that of cosmic wrinkles or any one of many artifacts—flowing from radiation from the galaxy or from the instrument—that looked like wrinkles. I felt the group's conviction rising that the emerging pattern we saw was indeed cosmic in origin, and my own confidence grew, too. Confidence can be treacherous, as it can erode one's determination to continue looking for errors in the data. We had to keep checking.

I went to Goddard immediately on returning from Antarctica in mid-December, stopping only briefly in Berkeley to exchange clothes. A meeting to review systematic errors had been scheduled for the end of the month, one last collective scrutiny of the data and methods of analysis, searching for anything that might be leading us astray, even at this late stage. Al Kogut had been managing the day-to-day systematic error studies and had done a magnificent job of maintaining vigilance. Al had independently repeated some of my work and tried new things from his fresh point of view. After a while the freshness faded and Al began to think he had uncovered every possible systematic error. That's when he and I began to argue. He would say something like, "Look, the systematic error limits are such and such. We have done all these wide-ranging tests. I've been through it five different times, and I can't imagine anything that would have gotten through and still caused a signal." And I would counter by warning, "Al, you're getting cocky. How can you know there isn't something out there that hasn't slipped through all our tests?"

Something had, as Al discovered shortly before the December meeting. The instrument partially remembered the prior measurement and included part of it in the next measurement. It was as if you photocopied a piece of paper on which the ink was still wet; part of the ink would stick to the glass and be recorded on the next copy. In our case, the result was a series of slight correlations between one measurement and the next. As a result, it produced faint patterns where none existed, creating the illusion of discrete regions a few degrees wide in the map that looked like wrinkles. The discovery of this effect was a salutary reminder that we had to keep our guard high.

The first day of the meeting included a presentation to NASA headquarters officials, to brief them on the progress of the COBE project. By this time the team had been working for two years on building maps of the sky. It had been a grinding time, the kind of experience that, in the absence of spectacular breakthroughs, all too easily saps motivation and enthusiasm. Team members had done superlatively everything that had been asked of them. We had met the criteria established for the success of the DMR—in terms of standards of data quality, collection, and analysis. I decided to make something of a ceremony of the announcement that we had met these criteria—known as the Level One specifications, in NASA parlance—and so turned up in a tuxedo. I declared I was proud of the team's dedication and that, formally, the project was matchless. (We did not tell the NASA officials what we could see on our maps—that would have to wait until we finished the systematic error review two days later.) The tuxedo was a big hit.

My attire that day represented a statement that the project had passed a major milestone: Until that point, we had operated on the premise that any fluctuation in the background radiation was a potential artifact of some kind, to be viewed with suspicion. Now it was time to devote significant resources to characterizing the structural features of the fluctuations, a picture of the universe three hundred thousand years after it formed.

Fifteen years earlier, back in 1977, we had been through a similar experience with the measurements of the cosmic background radiation we had obtained from the U-2-borne DMR. We were left with a uniform glow—no fluctuations in the background, no wrinkles in time. The COBE data were different, we were now certain of that. The new map showed a pattern of variation in the background radiation, no doubt about it. The question we were working toward was this: If the wrinkles are real, what range of sizes do they exist in, and how numerous is each class size? Demonstrating the reality of the wrinkles would bring us close to the reality of creation by the big bang; determining their size and number would give us a glimpse of how creation occurred.

Laurie Rokke continued to oversee data-processing operations at the COBE Data Analysis Center. Chuck Bennett oversaw the continuing flight operations, among other invaluable efforts. Al pushed on with the documentation of our systematic error studies. While back in Berkeley I developed additional software to analyze the data. Two graduate students, Luis Tenorio and

Charley Lineweaver, worked with me daily in this part of the effort, though we kept our work separate so as to provide checks on each other. Luis and I first looked carefully at the dipole signal to be sure each part of the data was correct. Watching the dipole emerge from the signals each time we applied the software analysis was a constant affirmation that we were on the right track. If each day's dipole did not look like the previous day's, we would know something was badly wrong, either with the DMR's collection of radiation or with the analysis. Each day the dipole emerged, with the same shape and same dimensions. Day by day it matched; and week by week. The only variation we saw was that caused by the motion of the Earth in its orbit around the Sun—confirming that Galileo was right.

Next we confirmed what we had been seeing for a year: that there was a real quadrupole present in the map. The shape of galactic emission is predominantly quadrupolar, so this made the galactic emission maps extremely important. If our galactic map was amiss in any way, the result would obscure the quadrupole. Giovanni's and my Antarctic efforts provided peace of mind. It was fine. Chuck Bennett, Giovanni De Amici, and Gary Hinshaw labored intensively checking and fitting the DMR data and external galactic maps to estimate the galactic contamination. The quadrupole signal is incredibly weak, less than six parts in a million—this is very close to noise, or foreground interference. But it is real. If the map is envisioned as an oval, lying horizontal, the dipole signal shows up as red and blue coloration at opposite poles of the oval. The quadrupole signal includes red and blue coloration at the top and bottom edges of the oval, giving four departures from the average background.

We were confident that the quadrupole was a real cosmic signal, a first hint of structure that was present in the fabric of the universe at three hundred thousand years after the big bang. How to explore the rest of the structure—if present—was the challenge.

Following our dictum of checking and double-checking as independently as possible, I had developed two programs to analyze the distribution of the number of the wrinkles versus their angular size in the sky. This is known as the power spectrum. The senior graduate student, Luis Tenorio, then took over the development, testing, and running of one of the programs—the spherical harmonic analysis. My newer graduate student, Charley Lineweaver, joined me on the second program, which sought correlation in temperature between different parts of the sky. Compared with spherical harmonic analysis, this method was less well known to most astrophysicists, and was a way of double-checking the results. I had one version of the program, and Charley a modified version, so we provided further checks for each other.

I worked as quickly as I could. Half of each day I spent checking on the status of other work at Berkeley and Goddard. The rest of the time, often late into the night, I concentrated on determining the power spectrum, in parallel with Luis and Charley. Luis and I ran into a problem looking past the dipole and quadrupole to the higher-number spherical harmonics. The large cut we had to make to ensure there was no contamination from the galaxy kept us from seeing the full pattern of the spherical harmonics across the sky. We had to extrapolate across the galactic plane region. With many spherical harmonics, how to fill in the blank region was ambiguous. It was apparent that the correlation method would be more reliable. Soon Charley and I had our programs running in tandem, so we were able to make analysis runs and compare results within a couple of days of each other. By late January and into early February, the results were beginning to gel, but they still didn't quite make sense. I tried all kinds of different approaches, plotting data in every format I could think of, including upside down and backward, just to try a new perspective and hoping for a breakthrough. Then I thought, why not throw out the quadrupole—the thing I'd been searching for all those years—and see if nature had put anything else there! I guessed this might be a way of determining whether our map was merely the result of noise, a cacophony of random radiation from many sources, or contained a genuine signal of wrinkles in more detail than the quadrupole. Working late into the night, I quickly modified my program to test the idea. That February evening I ran the new software and had a sense I was close to something important.

The search for the wrinkles had been a long endeavor, stretching back more than fifteen years; and the previous two years of the inexorable search for systematic errors inevitably precluded a Eureka-like experience. It was more like excavating a buried city, beginning by uncovering a first hint of interesting structure as the first layer of sediment is brushed aside. As further layers are removed, the form of the city's edifices begins to take shape, but slowly. Only when the final layers have been removed do we see the overall configuration of all the buildings. That evening I had the feeling that we were poised to sweep away the final obscuring layers in our quest for wrinkles.

All along I had been making simulations of how the wrinkles should look if the inflationary big bang theory was correct. There would be wrinkles of all sizes, from large to small, but with the same average area of sky occupied by each size class and the same average variation in amplitude—this is known as scale invariance. Inflation theory predicted such a size distribution, the products of quantum fluctuations at the instant of creation. This kind of size distribution of wrinkles would lead to the formation of structures of different sizes that we see in the universe today.

I ran the new program on these simulations as well as on our data. The results from the simulations were distinctive. And when I compared them with the results of subtracting the quadrupole from the map of the background radiation, it was clear that they matched the scale invariance predicted by inflation theory. The comparison was beautiful in an esoteric-modern-art way. Was I happy? Yes. Ecstatic, certainly. But not really surprised. By this time, my mind was so immersed in the data that the answer was obvious, if not consciously so. My immediate reaction was to shout it out—it looked so beautiful. A kaleidoscope of ideas was sorting itself out in my mind, making a perfect pattern. The big bang was correct; inflation theory worked; the pattern of the wrinkles was about right for structure formation by cold dark matter; and the size distribution would yield the major structures of today's universe, under gravitational collapse through 15 billion years.

The pattern of wrinkles I saw on the map was primordial—I knew it in my bones. Some of the structures represented by the wrinkles were so large that they could only have been generated at the birth of the universe, not later on. I was staring at primordial wrinkles in time, the imprint of creation and the seeds of the modern universe.

My exhilaration was extraordinary, but I *didn't* shout out. Even though I felt it was right, I still needed to be sure that I wasn't being fooled by some technical trick. Why, I puzzled, did I have to remove the quadrupole to see the wrinkles? I wanted Charley to check what I had done, using his system. Struggling to regain my composure, I penned a neutral note, suggesting Charley might like to modify his software to remove the quadrupole from the map, just to see what he might get. I then set off on the short trip down from the Lawrence Berkeley Laboratory to my home, gazing (unseeing) at the lights across San Francisco Bay. My mind was elsewhere, a turmoil of triumph tempered by concern. Celebrations—if appropriate—could come after my results had been checked.

Toward the end of the following day, I wandered into Charley's room and, as nonchalantly as I could, asked how the run had gone. He hadn't gotten around to doing it yet, he said. He had an idea about modifying the program—perhaps he could do better if we could get information on the data points nearest to zero separation. Hiding my frustration, I acknowledged that was an interesting idea, and again urged him to subtract the quadrupole. Do both things at once, I suggested.

Another day passed, and Charley was still distracted by his idea for improving the program. He still hadn't subtracted the quadrupole from the map. I could bear it no longer. About midnight I went across the hall to where he was working and said: "Don't go home until you run the maps through your program with the mean, the dipole, *and the quadrupole* subtracted." I made it sound urgent, but didn't say why, and then left for home. About three hours later, the run completed, Charley slipped some plots under my office door.

I arrived at my office early the next morning, because I'd arranged a conference call with the team at Goddard to discuss progress with the quadrupole. As I entered my office I almost stepped on the plots Charley had slipped under the door. Picking them up, I saw a white Post-it: "Here are the plots you asked for. Eureka?" He had found exactly what I'd seen. When he got into work late that day we went carefully through what each of us had done and tried a few more combinations of data. Then I sent out a note to Ned Wright and the DMR team members in Greenbelt, asking them to try the same approach.

Within days we were all getting the same result. My spirits soared, but there was still a lot to do. Iron out the details, make final checks, and prepare manuscripts for publication of our results in the *Astrophysical Journal*. That took the rest of February, March, and half of April. Gradually we began to understand why the quadrupole signal was so faint, almost at vanishing point. One problem was that our galaxy had a quadrupole of its own, opposite in direction to the cosmic signal. It was as if there were a cosmic conspiracy to minimize the signal, testing our ingenuity and patience in searching for it. In the face of adversity, we had triumphed and secured the Holy Grail of cosmology. This was the reward of long years of effort by many people: The COBE team, including the many managers, engineers, technical and other workers; the Science Working Group; and the various DMR teams—first, the splendid instrument builders, followed by the equally outstanding people who processed and analyzed the data. All in all, this was a huge team effort in which a myriad of things worked when they had to.

Two weeks before the American Physical Society meeting, where we planned to present six papers describing our results in detail, we held a science team meeting at Nancy Boggess's house, just outside Washington, D.C. It was a pretty upbeat affair. Nevertheless, not everyone was positive. Philip Lubin, my former graduate student from the U-2 days, had been collecting data at the same site we had used in Antarctica, looking for wrinkles at smaller angles. Phil was convinced that they saw no signals, suggesting what the DMR saw might be wrong. Stephan Meyer of MIT and Ed Cheng of Goddard also urged caution, but not as strongly as Phil. They had launched a radiometer aboard a balloon in the fall of 1989, about the same time as we launched COBE. Data analysis was causing a lot of trouble—I could sympathize with that—but they were beginning to conclude that there was no quadrupole, no pattern of wrinkles, to be seen.

Naturally, I should have preferred to hear that both teams

confirmed our results, and so I was somewhat disconcerted by the double challenge. It was especially discomforting to know that both teams planned to present their data at the Washington meeting, where their talks were scheduled to follow the DMR presentations. Would there be instant controversy, with the COBE team claiming to discover wrinkles, rapidly followed by two denials? I hoped not. I knew I wouldn't have an opportunity to see their data before the meeting, so I couldn't assess carefully how strong their counterclaims were. But by this time I was certain that our claim was very strong. We argued back and forth, and someone suggested perhaps we should wait before going ahead with our announcement. I believed we had been cautious long enough, and that we had taken all reasonable precautions to ensure that the signal we had was cosmic in origin. "We're going ahead—I am confident in our results," I asserted finally. "I stake my personal reputation on our results."

The night before the Physical Society meeting, Al, Giovanni, and I stayed late making copies of our papers to hand out at the conference. Then Giovanni and I went to my house to finish preparations for our talks. We made placards that would be props in my talk, and that members of the team would hold up at appropriate times. Each placard carried part of our results and was labeled L=0, L=1, L=2, and so on up to L=20. Each L represented a look at hot and cold regions separated by an angle 180°/L. In effect, each placard represented a more and more detailed scrutiny of the wrinkles. I planned to call on individuals by turn to hold up the placards as my talk progressed, describing first the background (L=0), then the dipole (L=1), then the quadrupole (L=2), and so on. It was a way of bringing all the team members in on the presentation.

By five the next morning we were packing boxes with papers and props, and then we set off for downtown Washington. We arrived at the auditorium at seven-thirty, to find it locked and no one around. By eight o'clock there were people everywhere—and a TV crew from WQED in Pittsburgh. It was a taste of things to come. Six of us from the COBE team were scheduled to talk: I was the lead-off speaker, followed by Gary Hinshaw, Chuck Bennett, Ned Wright, Al Kogut, and Giovanni De Amici. There was a lot to organize before events began, and I was largely unaware of the building excitement. I distributed the placards, made sure I knew where all the team members were, and waited to be called to the podium. I had been allotted twelve minutes.

"Good morning," I began. "I've a lot to say, and not much time, so I'll get right to it."

I planned to move toward the announcement in steps, because it was logical to do so and it added drama to the event. I was having some fun after not talking for so long. First I reminded the audience of the COBE result from two years earlier, that the cosmic background radiation matched the characteristics of blackbody radiation and therefore lent credence to big bang theory. I held up the placard labeled L=0, which showed the blackbody spectrum, and then I passed it to John Mather. Next, I said that the differential microwave radiometer had precisely measured the dipole, which we had discovered and I had announced exactly fifteen years earlier. I held up the next placard, labeled L=1, which showed the dipole: This demonstrated that the instrument was working properly. Then I passed it to Jon Aymon. Then, pausing just slightly for effect, I held up the next, L=2, and announced, "We have a quadrupole." The sense of relief in the auditorium was palpable, as everyone knew the import of what I was saying. I passed it down to Chuck Bennett, pointing out that he would say more about the quadrupole in his talk.

Even those in the audience who had guessed that our big announcement was going to be the discovery of the quadrupole did not anticipate that we had gone even further. I held up the L=3, L=4, and L=5 placards simultaneously, calling on the team members to hold up more placards, with ever-increasing L numbers, all the way up to L=20. That was clearly a shock to most people, as we revealed that not only had we found the quadrupole, which would have been momentous, but that we had also detected a spectrum of wrinkles of all sizes.

For me, the moment marked the culmination of an eighteen-year search, and for cosmology a major milestone on the long journey to understanding the nature of the universe. Very simply, the discovery of the wrinkles salvaged big bang theory at a time when detractors were attacking in increasing numbers. The result indicated that gravity could indeed have shaped today's universe from the tiny quantum fluctuations that occurred in the first fraction of a second after creation. When Stephen Hawking later commented that we had made "the most important discovery of the century, if not of all time," he may have been exaggerating, but it was still momentous. Before the discovery, our understanding of the origin and history of the universe rested on four major observations: first, the darkness of the night sky; second, the composition of the elements, with the great preponderance of hydrogen and helium over the heavier elements; third, the expansion of the universe; fourth, the existence of the cosmic background radiation, the afterglow of a fiery creation.

The discovery of the wrinkles that were present in the fabric of time at three hundred thousand years after creation becomes the fifth pillar in this intellectual edifice and gives us a way of understanding how structures of all sizes, from galaxies to superclusters, could have formed as the universe evolved during the past 15 billion years.

The evolution of the universe is effectively the change in distribution of matter through time—moving from a virtual homogeneity in the early universe to a very lumpy universe today, with matter condensed as galaxies, clusters, superclusters, and even larger structures. We can view that evolution as a series of phase transitions, in which matter passes from one state to another under the influence of decreasing temperature (or energy). We are all familiar with the way that steam, on cooling, condenses: This is a phase transition from a gaseous to a liquid state. Reduce the temperature further, and eventually the water freezes, making a phase transition from the liquid to the solid state. In the same way, matter has gone through a series of phase transitions since the first instant of the big bang.

At a ten-millionth of a trillionth of a trillionth of a trillionth (10^{-42}) of a second after the big bang—the earliest moment about which we can sensibly talk, and then only with some suspension of disbelief—all the universe we can observe today was the tiniest fraction of the size of a proton. Space and time had only just begun. (Remember, the universe did not expand *into* existing space after the big bang; its expansion created space-time as it went.) The temperature at this point was a hundred million tril-

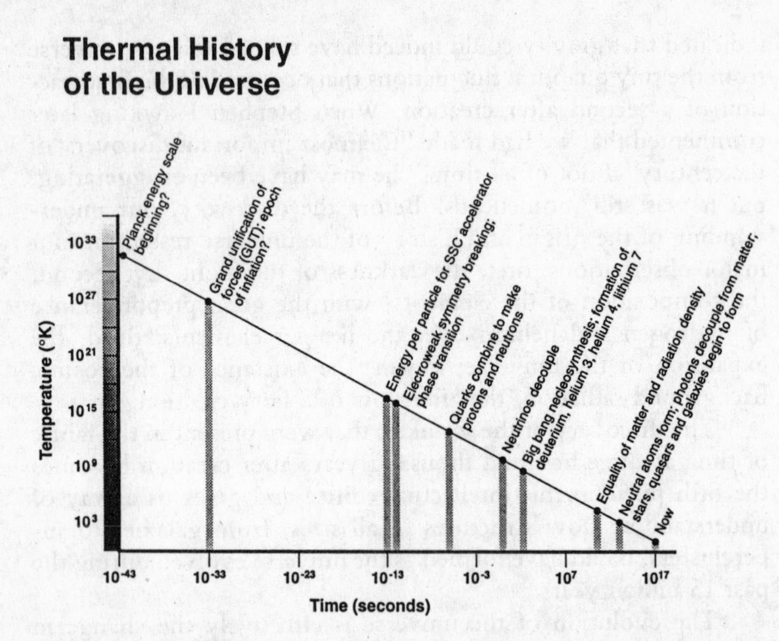

lion trillion (10^{32}) degrees, and the three forces of nature—electromagnetism and the strong and weak nuclear forces—were fused as one. Matter was undifferentiated from energy, and particles did not yet exist.

By a ten-billionth of a trillionth of a trillionth of a second (10^{-34} second) inflation had expanded the universe (at an accelerating rate) a million trillion trillion (10^{30}) times, and the temperature had fallen to below a billion billion billion (10^{27}) degrees. The strong nuclear force had separated, and matter underwent its first phase transition, existing now as quarks (the building blocks of protons and neutrons), electrons, and other fundamental particles.

The next phase transition occurred at a ten-thousandth of a second, when quarks began to bind together to form protons and neutrons (and antiprotons and antineutrons). Annihilations of particles of matter and antimatter began, eventually leaving a slight residue of matter. All the forces of nature were now separate.

The temperature had fallen sufficiently after about a minute to allow protons and neutrons to stick together when they collided, forming the nuclei of hydrogen and helium, the stuff of stars. This soup of matter and radiation, which initially was the density of water, continued expanding and cooling for another three hundred thousand years, but it was too energetic for electrons to stick to the hydrogen and helium nuclei to form atoms. The energetic photons existed in a frenzy of interactions with the particles in the soup. The photons could travel only a very short distance between interactions. The universe was essentially opaque.

When the temperature fell to about 3,000 degrees, at three hundred thousand years, a crucial further phase transition occurred. The photons were no longer energetic enough to dislodge electrons from around hydrogen and helium nuclei and so atoms of hydrogen and helium formed and stayed together. The photons no longer interacted with the electrons and were free to escape and travel great distances. With this decoupling of matter and radiation, the universe became transparent, and radiation streamed in all directions—to course through time as the cosmic background radiation we experience still. The radiation released at that instant gives us a snapshot of the distribution of matter within the universe at three hundred thousand years of age. Had all matter been distributed evenly, the fabric of space would have been smooth, and the interaction of photons with particles would have been homogeneous, resulting in a completely uniform cosmic background radiation. Our discovery of the wrinkles reveals that matter was not uniformly distributed, that it was already structured, thus forming the seeds out of which today's complex universe has grown.

Those regions of the universe with a higher concentration of matter exerted more gravitational attraction and therefore curved space positively; less dense areas had less gravitational attraction, resulting in less curvature of space. When radiation and matter decoupled three hundred thousand years later, the suddenly released flux of cosmic background photons bore the imprint of these distortions of space, showing the wrinkles we see in our maps: Radiation traveling from the denser areas looks cooler than the average background; that from less dense, warmer.

Matter in the universe, as we saw earlier, is of two kinds—dark matter and visible matter—and their role in gravitational formation of structure is different. Dark matter, which by its nature is unaffected by radiation but responsive to gravity, would have started forming structure much earlier than visible matter, which is buffeted by the energetic flux of photons. Molded by the contours of space that originated as quantum fluctuations in the inflationary universe, dark matter could have begun to aggregate, under the influence of gravity, as early as ten thousand years after the big bang. At three hundred thousand years, the decoupling of matter and radiation freed ordinary visible matter to be attracted to the structures formed by dark matter. As the visible matter aggregates, stars and galaxies form. A nice image here is the way cobwebs, often unseen in ordinary light, become strikingly visible when dew that settles on their strands during the night is lit by the morning sun. The gossamer network of galaxies we see in the night sky is the shimmering dew on a cosmic cobweb, as visible matter outlines the shape of structures of invisible dark matter, to which it has been drawn by gravitational attraction.

Because of the limitations of the differential microwave radiometer aboard COBE, the resolution of our maps is relatively poor. The smallest objects we can see as wrinkles are enormous, leading to structures as large as or larger than the Great Wall, a vast, concentrated sheet of galaxies stretching many hundreds of millions of light-years across. When we can achieve greater resolution, I expect us to be able to discern structures the size of galaxies. Despite current limitations, however, the message of our results—the message that engendered so much relief among cosmologists that April day—was clear. Fred Hoyle once claimed that big bang theory failed because it could not account for the early formation of galaxies. The COBE results prove him wrong. The existence of the wrinkles in time as we see them tell us that big bang theory, incorporating the effect of gravity, can explain not only the early formation of galaxies but also the aggregation within 15 billion years of the massive structures we know to be present in today's universe. This is a triumph for theory and observation.

We did not anticipate the extent of the reaction to our announcement, both immediately and in the subsequent weeks. I knew the occasion would be important for cosmologists. When Gary Hinshaw passed me on the way to the podium to give the talk after mine he muttered, "How am I supposed to follow *that*!" Later, when astrophysicist Sun Rhie rose to give her paper on topological defects, she opened by saying, "After the COBE announcement, I don't know if I should bother to give this talk." Just as our discovery supported gravitational accretion from primordial fluctuations and inflation, it undermined competing theories, particularly the theory of topological defects. This theory predicted a very different pattern of size and number of wrinkles from what we observed, and so was falsified. David Spergel, coauthor with Neil Turok of the textures version of topological defect theory, declared after the session, "We're dead." Paul Steinhardt, a leading inflation theorist, said, "I'm glad I was here for this historic day. . . . I wish I had brought my students."

This was gratifying, of course, particularly when I was finally able to see the presentations by the MIT and South Pole teams. Although Lyman Page described the MIT data as contradicting COBE's, I could see from his final viewgraph that wasn't the case. They claimed there were no wrinkles, but it was clear to me that the correlation curve was consistent with COBE's data.* A wave of relief passed over me as I realized there would be no controversy over this. And when Todd Gaier gave the South Pole results with his supervisor, Phil Lubin, looking on, they came close to disagreeing with us, but there were many possible explanations for the discrepancies. We were home free—for that day, at any rate.

I had a few moments to grab a sandwich, think about the press conference scheduled at noon, and search for the viewgraph that we had made up to explain our results to the press. It was delivered at the last moment fresh from the drawing board.

Then we walked down to the room where the press conference was to be held. There was a huge crowd there; in the back of the room a row of television cameras and a bank of lights, making me blink with more than just surprise. As Ned Wright,

*In December 1992, MIT team member Stephan Meyer reported to the Texas-Pascos meeting in Berkeley that their balloon work has essentially confirmed the COBE DMR results.

Chuck Bennett, Al Kogut, and I got up on the podium, I started to realize the immense public interest in our new results. I was still composing myself to express now, both to the press and the television simultaneously, the meaning of our results in words and metaphors that would be readily understandable.

I was first to speak and tried to present the high-level facts and significance of what we had done: "We had observed the oldest and largest structures ever seen in the early universe," I began. "These were the primordial seeds of modern-day structures such as galaxies, clusters of galaxies, and so on. Not only that, but they represented huge ripples in the fabric of space-time left from the creation period." Many other questions were raised, but primarily they centered on two issues: How big were the

At the April 23, 1992, press conference, George Smoot, holding a view graph, announces wrinkles in space-time. AP/WIDE WORLD PHOTOS

structures? and, What was the significance of our results?

I answered that superlatives failed to convey a real sense of these structures, which stretched across such vast distances that they dwarfed the Great Wall, itself hundreds of millions of light-years in extent. Some were so large that light had not yet traveled across them in the age of the universe. How fundamental? I gave many comparisons, but the one picked up most by the press was: "If you're religious, it's like seeing God." The big bang is a cultural icon, a scientific explanation of the creation. Out of forty minutes of questions from newspaper and television journalists, with repeated requests to rephrase what the COBE results meant to laypeople, that single line was the most quoted and remembered.

In cosmology there is a confluence of physics, metaphysics, and philosophy—when inquiry approaches the ultimate question of our existence, the lines between them inevitably become blurred. Einstein, who was devoted to a rational explanation of the world, once said: "I want to know how God created the world. I want to know his thoughts." He meant it metaphorically, as a measure of the profundity of his quest. My own much-quoted remark was cast in the same mold.

Metaphorically meant or not, my remarks, and the comments of other COBE team members and other cosmologists, appeared in newspapers around the world, affirming the deep public interest in understanding the origin of the universe and our place in it. Ned Wright spoke after me, explaining in more detail what our results meant in terms of theories of structure formation and the need for dark matter. Chuck Bennett gave a brief statement about the results and interpretation and answered questions. Finally, Al Kogut reviewed the efforts we had taken to process the data carefully and ensure that we had removed all instrumental effects. Al explained the great difficulty of the experiment. After Al finished, the conference was opened to general questions. The press conference went on and on until finally it was called to a halt after a couple of hours—whereupon it broke up into small knots of reporters interviewing us, other COBE team members, and people like Alan Guth and Paul Steinhardt, authors of inflation theory.

In a while Philip Schewe and the other American Physical

Society press people led us back to the conference press office so that we could take calls and interviews from reporters not present. They set up six of us around the room to do radio, phone, TV, newspapers, and magazine interviews, and it went on with no letup through the afternoon into the evening. The only brief event I remember was around 4:30 or 5:00 P.M. I finished an interview and called across the noisy room to Philip Schewe, "Do you have press-piranha feedings like this very often?" Philip answered, "The only time I have ever seen it like this was for the cold fusion announcement." Suddenly, for a moment, the whole room was quiet. Everyone seemed to be thinking, What if this is wrong? What if we made a mistake? Then it was back to ringing phones and interviews.

Various COBE team members and other cosmologists were on TV, radio talk shows, and in newspapers for several days. The publicity and tremendous public interest provided a unique opportunity to discuss science with a very large audience and to promote the power of human endeavor in pursuing the mysteries of nature. I recognized from the experience people's profound hunger for the metaphors of creation, even if the science itself is intellectually taxing.

A powerful conviction for me, and one that I believe encourages confidence that one day we will understand the very essence of creation, is the idea that as we converge on the moment of creation, the constituents and laws of the universe become ever simpler. A useful analogy here is life itself, or, more simply, a single human being. Each of us is a vastly complex entity, assembled from many different tissues and capable of countless behaviors and thoughts. Trace that person back through his or her life, back beyond birth and finally to the moment of fertilization of a single ovum by a single sperm. The individual becomes ever simpler, ultimately encapsulated as information encoded in DNA in a set of chromosomes. The development that gradually transforms a DNA code into a mature individual is an unfolding, a complexification, as the information in the DNA is translated and manifested through many stages of life. So, I believe, it is with the universe. We can see how very complex the universe is now, and we are part of that complexity.

Cosmology—through the marriage of astrophysics and particle physics—is showing us that this complexity flowed from a deep simplicity as matter metamorphosed through a series of phase transitions. Travel back in time through those phase transitions, and we see an ever-greater simplicity and symmetry, with the fusion of the fundamental forces of nature and the transformation of particles to ever-more fundamental components. Go back further and we reach a point when the universe was nearly an infinitely tiny, infinitely dense concentration of energy, a fragment of primordial space-time. This increasing simplicity and symmetry of the universe as we near the point of creation gives me hope that we can understand the universe using the powers of reason and philosophy. The universe would then be comprehensible, as Einstein had yearned.

Go back further still, beyond the moment of creation—what then? What was there before the big bang? What was there before time began? Facing this, the ultimate question, challenges our faith in the power of science to find explanations of nature. The existence of a singularity—in this case the given, unique state from which the universe emerged—is anathema to science, because it is beyond explanation. There can be no answer to *why* such a state existed. Is this, then, where scientific explanation breaks down and God takes over, the artificer of that singularity, that initial simplicity? The astrophysicist Robert Jastrow, in his book *God and the Astronomers,* described such a prospect as the scientist's nightmare: "He has scaled the mountains of ignorance; he is about to conquer the highest peak; as he pulls himself over the final rock, he is greeted by a band of theologians who have been sitting there for centuries."

Cosmologists have long struggled to avoid this bad dream by seeking explanations of the universe that avoid the necessity of a beginning. Einstein, remember, refused to believe the implication of his own equations—that the universe is expanding and therefore must have had a beginning—and invented the cosmological constant to avoid it. Only when Einstein saw Hubble's observations of an expanding universe could he bring himself to believe his equations. For many proponents of the steady state theory, one of its attractions was its provision that the universe had no beginning and no end, and therefore required no explanation of what existed before time = 0. It was known as the perfect cosmological principle.

A decade ago Stephen Hawking and Jim Hartle tried to resolve the challenge differently, by arguing the singularity out of existence. Flowing from an attempt at a theory of quantum gravity, they agreed that time is finite, but without a beginning. This is not as bizarre as it sounds, if you think of the surface of a sphere. The surface is finite, but it has no beginning or end—you can trace your finger over it continuously, perhaps finishing up where you began. Suppose the universe is a sphere of space-time. Travel around the surface, and again you may finish up where you started both in space and time. This, of course, requires time travel, in violation of Mach's principle. But the world of quantum mechanics, with its uncertainty principle, is an alien place in which otherworldly things can happen. It is so foreign a place that it may even be beyond human understanding, children as we are of a world of classical Newtonian mechanics.

We simply do not know yet whether there was a beginning of the universe, and so the origin of space-time remains in terra incognita. No question is more fundamental or more magical, whether cast in scientific or theological terms. My conviction—perhaps I should say my faith—is that science will continue to move ever closer to the moment of creation, facilitated by the ever-greater simplicity we find there. Some physicists argue that matter is ultimately reducible to pointlike objects with certain intrinsic properties. Others argue that fundamental particles are extraordinarily tiny strings that vibrate to produce their properties. Either way, in combination with certain concepts such as inflation, it is possible to envisage creation of the universe from almost nothing—not nothing, but practically nothing. Almost creation *ex nihilo*, but not quite. That would be a great intellectual achievement, but it may still leave us with a limit to how far scientific inquiry can go, finishing with a description of the singularity, but not an explanation of it.

To an engineer, the difference between nothing and practically nothing might be close enough. To a scientist and certainly to a philosopher, such a difference, however minuscule, would be everything. We might find ourselves experiencing Jastrow's bad dream, facing a final question: Why? "Why" questions are not amenable to scientific inquiry and will always reside within philosophy and theology, which may provide solace if not material explication.

But what if the universe we see were the only one possible, the product of a singular initial state shaped by singular laws of nature? By now it is clear that the minutest variation in the value of a series of fundamental properties of the universe would have resulted in no universe at all, or at least a very alien universe. For instance, if the strong nuclear force had been slightly weaker, the universe would have been composed of hydrogen only; slightly stronger, and all the hydrogen would have been converted to helium. Slight variation in the excess of protons over antiprotons—one billion and one to one billion—might have produced a universe with no baryonic matter or a cataclysmic plenitude of it. Had the expansion rate of the universe one second after the big bang been smaller by one part in a hundred thousand trillion, the universe would have recollapsed long ago. An expansion more rapid by one part in a million would have excluded the formation of stars and planets.

The list of cosmic coincidences required for our existence in this universe is long, moving Stephen Hawking to remark that "the odds against a universe like ours emerging out of something like the big bang are enormous." Princeton physicist Freeman Dyson went further, and said: "The more I examine the universe and the details of its architecture, the more evidence I find that the universe in some sense must have known we were coming." This concatenation of coincidences required for our presence in this universe has been termed the anthropic principle. In fact, it is merely a statement of the obvious: Had things been different, we would not exist. It may be that many different universes are possible, and many may exist in parallel with our own. Inflation theory can be interpreted in this way, with our universe budding off a larger fabric of space-time—like one strawberry in a patch of many strawberries. My speculation, however, is that because things become simpler as we near the moment of creation, there was only a limited range of possibilities; indeed, perhaps only one, with everything so perfect that it could have been no other way.

In this case, what could we say about the ultimate question?

Big Bang:
The Expanding and Evolving Universe

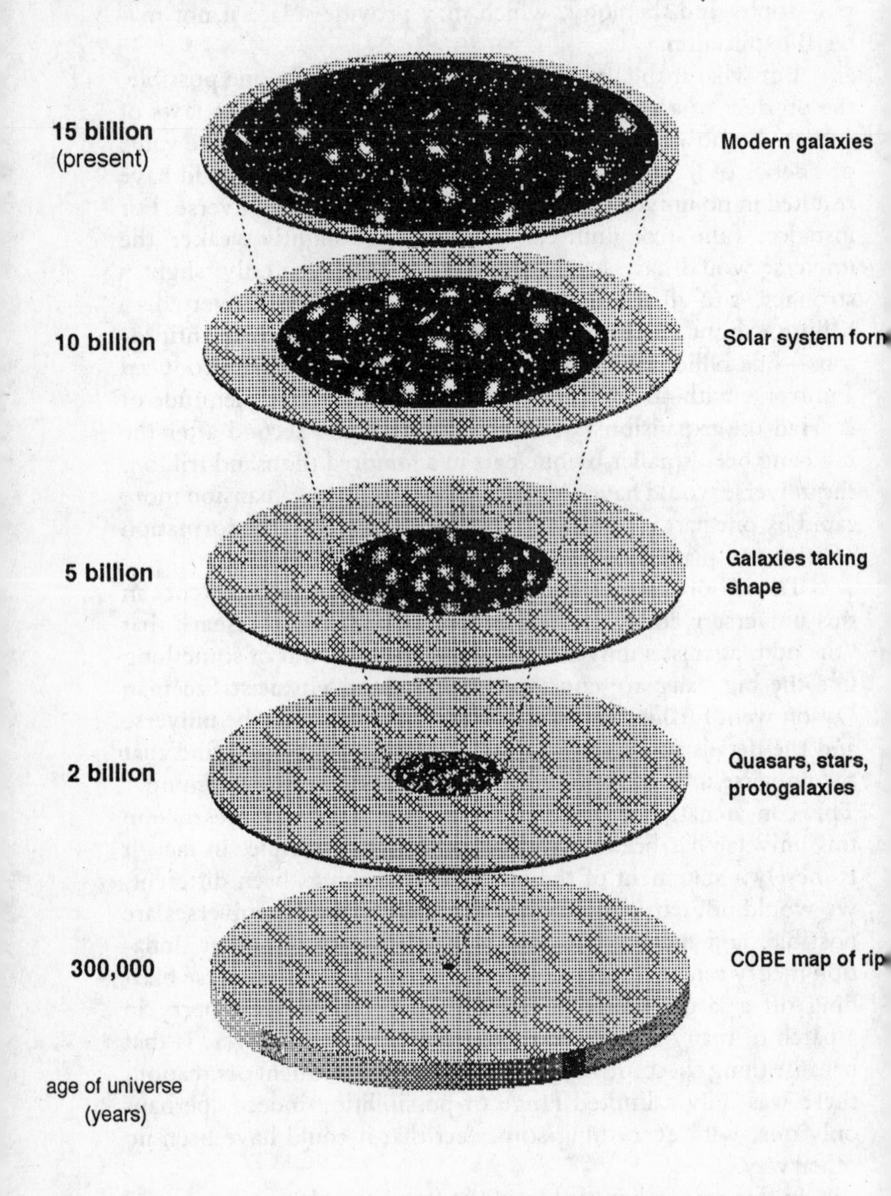

That God had no choice in how the universe would be, and therefore need not exist? Or that God was very smart, and got it just right? In any case, science would still be left contemplating the question: Why these conditions and not others? Or perhaps the comprehensibility of the universe in these terms is sufficient explanation. The truth and treasure of the universe is its own existence, and our quest for that truth and treasure will be eternal, like the universe itself.

Our discovery of the wrinkles in the fabric of time is part of that eternal quest and marks an important step forward in this golden age of cosmology. Suddenly, pieces of a larger puzzle begin to fall together: Inflation looks stronger, and dark matter more real. Our faith in the big bang is revitalized: To the dark night sky, the composition of the elements, the evidence of an expanding universe, and the afterglow of creation is added a means by which the structures of today's universe could have formed. The creativity of the universe is its most potent force, forming through time the matter and structures of stars and galaxies, and, ultimately, us. The wrinkles are the core of that creativity, assembling structure from homogeneity.

The quest will continue, with the dual goals of discovering dark matter and understanding the origin of space-time. No one knows where the answers will come from, but if recent history is any guide, a combination of observational science and particle physics will be important. More satellite ventures are planned, and the superconducting supercollider (SSC), one of the boldest scientific endeavors of all time, is under construction in Texas. Without doubt it is expensive—perhaps as much as ten billion dollars. But the history of this science has been driven by the development of instrumentation. There may be no other way of understanding the first instants of creation than by re-creating big bangs of sorts in a massive machine like the fifty-four-mile circumference supercollider. As a culture, we have to decide whether the intellectual quest toward the ultimate question is worthwhile. The more we know about the history of the universe, the more we know about ourselves and the questions we are driven to ask.

In 1977, Steven Weinberg published *The First Three Minutes*, one of the finest popular books on cosmology ever written, and justly still in print. His book was based on a course he taught about gravitation and cosmology at MIT while I was a graduate student there. His class influenced my decision to enter cosmological research. Toward the end of his book, Weinberg muses on the questions we ask ourselves, particularly the conviction that, somehow, humans are not a mere cosmic accident, the chance outcome of a concatenation of physical processes in a universe that dwarfs us on every scale. He expresses his view on the matter this way: "It is very hard to realize that [this beautiful Earth] is all just a tiny part of an overwhelmingly hostile universe. It is even harder to realize that this present universe has evolved from an unspeakably unfamiliar early condition, and faces a future extinction of endless cold or intolerable heat. The more the universe seems comprehensible, the more it also seems pointless."

I must disagree with my old teacher. To me the universe seems quite the opposite of pointless. It seems that the more we learn, the more we see how it all fits together—how there is an underlying unity to the sea of matter and stars and galaxies that surround us. Likewise, as we study the universe as a whole, we realize that the "microcosm" and the "macrocosm" are, increasingly, the same subject. By unifying them, we are learning that nature is as it is not because it is the chance consequence of a random series of meaningless events; quite the opposite. More and more, the universe appears to be as it is because it *must* be that way; its evolution was written in its beginnings—in its cosmic DNA, if you will. There is a clear order to the evolution of the universe, moving from simplicity and symmetry to greater complexity and structure. As time passes, simple components coalesce into more sophisticated building blocks spawning a richer, more diverse environment. Accidents and chance, in fact, are essential in developing the overall richness of the universe. In that sense (although not in the sense of quantum physics), Einstein had the right idea: God does not play dice with the universe. Though individual events happen as a matter of chance, there is an overall inevitability to the development of sophisticated complex systems. The development of beings capable of questioning and understanding the universe seems quite natural. I would be quite surprised if such intelligence has not arisen many places in our very large universe.

As I travel the world, I love to visit great art museums, to see classic sculptures, the works carved and painted and assembled by centuries of aesthetic visionaries. Cosmologists and artists have much in common: Both seek beauty, one in the sky and the other on canvas or in stone. When a cosmologist perceives how the laws and principles of the cosmos begin to fit together, how they are intertwined, how they display a symmetry that ancient mythologies reserved for their gods—indeed, how they imply that the universe *must* be expanding, *must* be flat, *must* be all that it is—then he or she perceives pure, unadulterated beauty.

The religious concept of creation flows from a sense of wonder at the existence of the universe and our place in it. The scientific concept of creation encompasses no less a sense of wonder: We are awed by the ultimate simplicity and power of the creativity of physical nature—and by its beauty on all scales.

Appendix:

Contributors to COBE

INDIVIDUALS WHO WORKED ON COBE

Eve Abrams
Robert Abresch
Rosa E. Acevedo
Kenton Ackerman
Norman Ackerman
Mario H. Acuna
Charles Adams
Kenton Adams
Susan E. Adams
William Adams
Mary Adkins
Darlene Ahalt
Eliezer Aharonovich
Angela Ahearn
Anisa Ahmad
James E. Akers
Cheryl Albert
Charles Alecknavage
Robert Aleman
Steve Alexander
Bruce Allen
Douglas Allen
Calvin Allison
Mary Aloupis
Thomas Amacher
David Amason
Wendy Ames
Julie Anderson
Matthew Anderson
Michael Anderson
Anthony Andoll
Leonard C. Andreozzi
Earl Angulo
William Anonsen
Carol Archambeault
Mark Arend
Nerses M. Armani
Robert M. Armstrong
Richard A. Arnold
Petar Arsenovic
Jorge A. Aviles
Lois Aylor
Paul Aylor
Jon Aymon
Sabinus Azoro
Karin Babst
Neal D. Bachtell
Charles Backus
Hortense Backus
Yoon Bae
Leon Bailey
Susan Bailey

Audrey J. Baker
Donald Baker
Barry Baltozer
Peter M. Baltzell
Anthony Banday
Melvin Banks
Dale F. Bankus
Quinton Barker
Joseph Barksdale
Jeffrey Barnes
William Barnes
Richard D. Barney
Joseph Baros
Andrew G. Barr
William Barrett
Harold S. Barsch
D. E. Bartels
Michael Barthelmy
Shivanand Basappa
Lisa Basiley
Gayle Bates
John G. Bauernschmidt
Mark A. Baugh
Robert Baumann
Carl Bayne
Phyllis Bayne
David Bazell
Earl Beard
Norman R. Beard
Renate Beaver
John M. Beckham
Richard D. Beckman
Howard Becraft
Reuven Bekhokhma
Anthony J. Bello
Porifirio Beltran
Sheela Belur
Jeannette Benavides
Robert Bender
Bud Bengtson
Matt Bengtson
Charles L. Bennett
Jerome Bennett
John Bennett
Joseph Bentley
Darlene Bently
Ronald L. Bento
Erik C. Berger
Franz Berlin
Graham Berriman
Colman W. Beulah
Naren Bewtra
John Bichell
Arlene R. Bigel
Sandra Biggs
Jeffery Blackwell
Diane Blair
Wilfredo Blance
Roycee A. Bland
Thomas Blaser
Michael Blizzard
Nathan Block
Garcia Blount
Diane Bobak
John Boeckel
Edward Boggess
Nancy W. Boggess
Jerry T. Bonnell
Carol Boquist
Francesco Bordi
Patricia L. Bordi
Richard Bost
Rene A. Boucarut
Richard Bouchard
David Bouler
Robert Bounds
Stephen Bove
John Bowden
Donald Bower
Gordon Bowers
Gregory Bowers
William T. Bowers
Priscilla Bowes
Jeffrey Bowser
Robert F. Boyle
Barbara Bradtmueller
Christine Brandford
Regis Brekosky
Keith Brenza
Polly Bresnahan

Carrie Brezinski
Vincent J. Briani
Mildred Brice
Ronald D. Brickerd
Donald E. Briel
Thomas C. Briggs
Lawrence Bromery
Marvin Brooks
James J. Brophy
Cecilia Brown
Heather Brown
Kimberly Brown
Leonard Brown
Luther Brown
Mark A. Brown
Stacey L. Broyles
Nathan Bruce
Calvin Bryant
James Buckeridge
David Buckholtz
Spike Bukowski
James Bullock
Frances Bunevitch
Christopher Bunyea
Harold Burdick
Shawn Burdick
William M. Burgess
Carol Burke
Raymond Burkhardt
Lynette Burley
Peter Burr
Valorie Burr
Douglas Burritt
Helen Burritt
Joseph I. Burt
Scott Burtis
Gretchen Burton
Frank Bush
John Busse
Thomas Butash
James Byrd
Jerome Byrd
Wallace D. Cacho
Robert W. Cade
James Caldwell
Jody Caldwell
William Campbell
Malcolm Cannon
Vince Canali
Ciro Cancro
George W. Cantrell
Dawn Canty
Darlene Capone
Clinton Carle
Hawk Carnahan
Armen Caroglanian
Nancy Carosso
Daniel Carrigan
Donald Carson
Karen Carter
Denise Cartwright
Enrico Caruso
Paul Caruso
Ronald Cary
Christopher A. Cascia
William Case
Michael K. Cassidy
Stephen H. Castles
Joseph Cauley
Carline J. Cazeau
Jose R. Cerrato
Robert Chalmers
William Chambers
Carolyn Chandler
Florence Chandler
Po Lam Chang
Susan Chang
Chim Chao
Hwai-Soon Chen
Edward S. Cheng
Wendy Cheng
Sarada Chintala
John Chitwood
Sang Choi
Jerry Christian
Kyle G. Christoff
Paul T. Christoff, Jr.
Harry Chrysohidis
Quoc Chung
Se Chung

Juan Cifuentes
Larry Clairmont
David Clark
John Clarke
Carroll Clatterbuck
Kenneth Claytor
Steve Clemenson
James Cleveland
Lawrence Cline
Wanda Clingerman
Joel Cohen
Robert Coladonato
Alexander M. Coleman
Maude Coleman
Melanie Coleman
Roy Coleman
David Collier
Ernest Collins
Joseph Colony
Jack Commerford
Kevin Connell
Michael W. Connor
Richard A. Conte
Eric Cook
Lawrence Cook, Jr.
Jan Cooksey
Carollyn Cooley
Gerald Cooper
Nabil Copty
Julie Corry
Kenneth C. Cory
Donald Cosner
David Cottingham
Jim Cottrell
Jack Coulson
Grady Cox
Stephen R. Cox
Carole A. Cramer
Eric Cramer
Allen J. Crane
Melvin Crompton
Donald F. Crosby
Leigh Croteau
John Crow
Patrick Cudmore
Donald D. Culin
David W. Culver
Carlos Cumberbatch
Robert Cummings
Herbert Cunningham
Ada Curtis
Angela Curtis
Cynthia Curtis
Omar S. Custodio
William Daffer
Terry Daley
Lisa Dallas
Christopher Daly
Richard Dame
Charles Dan
Anthony Danks
Nicholas D'Apice
David Dargo
Henry H. Darin
Donald D. Davis
Glen Davis
Lloyd D. Davis
Mitchell Davis
Giovanni De Amici
Robert DeFazio
Wallace DeGrange
Donald Deibler
Casey Dekramer
Thomas Delaney
Ronald C. Delbrook
Robert Denhardt
Marion DePriest
Larry W. Derouin
Rebecca J. Derro
Ashok Desai
Michael Devett
Edward Devine
Jennell Dewitt
John DiBartolo
James Dieter
Donald Diggs
Deborah A. Dionisio
Michael Dipirro
Elaine C. Dixon
Ronald L. Dixon

James Dodd
Leon Donde
Martin J. Donohoe
Robert Dorn
Hal Dougherty
Andreas Doulaveris
Ernst Doutrich
Dewey Dove
Gully Dowdy
Maurice Dowdy
William Doyden
Theodore Drabczyk
Robert Drollinger
Donna Drummond
Michael Drury
Joseph Ducosin
Cedric Duffield
Thomas Dugan
Betty Dumas
Larry Dumonchelle
Clyde E. Duncan
Anthony Dunston
Jeff Durachta
Dale Durbin
Eliahu Dwek
James Dye
Edward Dyer
Derrick A. Early
George E. Eastman
James Edelmann
Jerome Edwards
William Eichhorn
R. Einertson
Wayne Eklund
Dean Elliott
Mathew Elliott
Daniel Ely
Joseph Emmons
Roger Emralino
Robert D. Endres
Charles England
Gene Eplee
Robert Eplee
Mark E. Erickson
Thomas Erickson
Irven Errera
Steven Etchison
Karen Evans
William Evans
John Ewing
Randee Exler
Carin Facchina
Donald Fadler
Roger E. Farley
Frank H. Fash
Matthew J. Fatig
Michael W. Fatig
Francis Federline
Richard Fedorchak
Paul Feinberg
Irene Ferber
Michael Femino
Casimir A. Ferenc
Nelson J. Ferragut
Paula Ferring
Donald Field
Orlando Figueroa
Debra L. Filson
Kelvin Finneyfrock
Nova D. Fisher
Michael W. Fitzmaurice
Dale Fixsen
William H. Flaherty
Mark Flanegan
Lisa Flemming
Catherine D. Fleshman
Yury Flom
Jose Florez
Manuel Florez
Robert Flynn
Everett Fogle
Walter Folz
James Fonner
David Ford
Mary Ford
James Fournier
Billy G. Fox
Rudolph L. Foxwell
Anthony D. Fragomeni
Marcia Frances

Richard A. Franchek
Gene Francis
Joan Frank
Bryan Franz
Gregory Frazier
Richard Freburger
Martin E. Frederick
Immanuel Freedman
Richard Freeman
Thomas French
Henry Freudenreich
Robert Frey
Patricia R. Friedberg
Scott Friedman
Joan L. Fritz
Glen Fuller
Edwin H. Fung
Ronald R. Gagne
Joel Gales
Gur Gallent
John Galloway
Kevin Galuk
Frederick W. Gams
Kenneth Gardner
Christopher J. Garner
James Gary
Raul Garza-Robles
James Gatlin
Michael Gauss
Joseph Gauvreau
Gary Gavigan
Richard Gayo
John B. Geagan
Wayne Geer
Timothy C. Gehringer
Guy Germana
Earl C. Gernert
Frederick C. Gessner
Nardinder Ghuman
Judith Gibbon
John Gibson
Carl Gieger
Phil Gill
Jeanette Gilleland
David Gilman
Carey Gire
Richard H. Gladfelter
Edward L.Glenn
Jenny M. Glenn
Kevin R. Glenn
Elva Glover
Gene Gochar
Meng Goh
James A. Golden
Carlos Gomez-Rosa
Nilo Gonzales
Lawrence L. Good
Walter Goodale
Cassandra Nix Goodall
Sebert Goodall
Wayne Goodman
William Goodyear
Diana Gordon
Krys Gorski
William Gotthardt
Ira Graffman
Randall Graham
Angela C. Grant
David Gray
Richard Gray
Alexander Green
Charles Green
Walter L. Green
Barry N. Greenberg
Thomas Greenhorn
Linda Greenslade
Jeffrey Greenwell
Ronald C. Greer
Wayne Greer
Jerry M. Greyerbiehl
George Griffin
Robert Grigsby
Celine Groden
Daniel Grogan
Frederick C. Gross
Michael R. Gross
Thomas Grubb
Kurt Grubler
Samuel Gulkis
Jeffery S. Gum

Ved K. Gupta
Nancy Gusack
Paul Guy
John J. Habert, Jr.
Thomas R. Hackerwelder
Robert Hackley
Carl Haehner
Kathleen K. Hagan
Joseph Hage-Dagher
Sheila Haghighat
John Hagopian
Paul Haight
William B. Haile
Thomas P. Hait
Milton Halem
Gardeania Hamilton
Joseph Hammerbacker
Jeffrey J. Hampton
Ronald Hand
Paul Haney
Hagop Hannanian
Richard V. Hanold
William Happel
Thomas Harbach
Richard Harner
Abigail D. Harper
Steven Harper
Cheryl Harrington
Benjamin Harris
Charles A. Harris
Gary Harris
Horace Harris
Russell E. Harrison
Jon Harzer
Lois Haskins
Peter Hatfalvi
Michael G. Hauser
James Heaney
David Heckle
Robert D. Heidenreich
Jeffrey Hein
Doug Heine
Brian N. Helland
Hamid Hemmati
Edward Hemminger
Gerald Hempfling
John Henegar
Paul Henley, Sr.
John Henninger
David Hepler
Roberto P. Hernandez
Leon Herreid
Michael J. Hersh
Donald Hershfeld
Jules Hershfield
Thomas Heslin
Dale Hetrick
Patricia Hettenhouser
Tilak Hewagama
Kevin A. Hicks
Stanley C. Hilinski
James Hill
Mark L. Hill
Paul D. Hill
Roland Hill
Thomas Hill
Lawrence M. Hilliard
R. Kenneth Hinkle
Noel Hinners
Gary Hinshaw
George Hinshelwood
Triem Hoang
Mary Hoban
Richard Hockensmith
Janet V. Hodges
Henry Hoffman
Richard Hoffman
William Hoggard
Howard Holbrook
Neil Holby
George Holland
Sue Holland
Wayne Holland
Richard Hollenhorst
Cecelia Hollis
Michael Hollis
John Holloman
David Hon
Michael A. Honaker
John Hopkins

Richard A. Hopkins
Lynn Hoppell
Lee Horning
Charles E. Houchins
Joseph K. Howard
Elmer T. Howell
Qihui Huang
Alton Hubbard
Thomas Huber
David Huff
Allen D. Huffman
Richard Hughes
Wayne Hughes
Peter Hui
Edward Hulbert
Carolyn Humphrey
Patrick W. Humphrey
Ronald Hunkeler
Cleophus T. Hunt
Darlene Hunt
Russell Hurlbert
Jamil Hussain
Mark Hymowitz
Dominick Iascone
Joseph Iffrig
Mary Igal
William Iliff
Grace Incoom
Jeffrey L. Ingream
Elaine Inskeep
Sandra Irish
Richard Isaacman
Harold Isenberg
Murry Itkin
Patrick Izzo
Bud Jackson
Mattie Jackson
Patricia M. Jackson
Peter Jackson
Robert Jackson
Willard E. Jackson, Jr.
William Jackson
Daniel Jacobs
Atul Jai
Calvin L. James
John C. James III
Joseph James
Paul James
Rita Jankowski
Michael Janssen
William Jarrell
Elizabeth M. Jay
Jane Jellison
Diane Jenkins
Eric Jenkins
Linda Jenkins
Kirk Jennings
Kenneth Jensen
Joel Jermakian
Jay Jett
Barbara Johnson
Bernard Johnson
Carlton Johnson
Clarence Johnson
Helen Johnson
Leslie Johnson
Marva Johnson
Michael Johnson
Tamra T. Johnson
Thomas Johnson
William Jolly, Jr.
Anthony Jones
Carol Jones
David Jones
Frank Jones
Isaiah Jones
Maggie E. Jones
Morris Jones
Richard M. Jones, Jr.
Shirley Jones
Willie Jones
Shawpin Jong
Carolyn Judd
Edward Kaita
Ford Kalil
Susan Kaltenbaugh
George Kambouris
Allan L. Kane
Lisa Kane
K.S. Kannon

Linda Drexler Kanzler
Keith Kappes
Bernard F. Karmilowicz
Gabriel Karpati
Charles Katz
Lonny R. Kauder
Marvin Kaufman
Joseph Keating
Ronald Kedzierski
Phillip Keegstra
Michael E. Kefauver
Daniel Keith
Amy Kekeisen
Carl W. Kellenbenz
Michael W. Keller
John Kelly
Linda Kelly
William Kelly
Thomas Kelsall
Myron Kemerer
Steven Kempler
Cherry A. Kenney
Richard Kennon
Peter Kenny
James Kerley
Deborah D. Kershner
Robert A. Kichak
Patrick L. Kilroy
Sang Kim
Charles W. King
James S. King
Henry Kingwood
Robert Kinwell
Laurie Kipple
Frank Kirchman
James D. Kirk
Donald Kirkpatrick
Robert Kirkpatrick
Emil R. Kirwan
Francis J. Kisner
James Kistler
Jan Kjellberg
Bernard J. Klein
Herman Kleiner
Daniel A. Klinglesmith
Douglas E. Knight
Lyle Knight
Dawn Knode
John M. Koenig
Alan Kogut
Leonid Kokourine
Diane Kolos
Andrew Korb
Carl Kotecki
Robert Kozon
Betty Kramer
William Krause
Ronald Krellen
Michael Kridler
James Krise
George Kronmiller
Donald Krueger
Timothy O. Krueger
Peter Kryszak
Michael Kulpak
Carol Kulwich
Vijaya Kumar
Robert Kummerer
Charles Kung
Michael Kurtz
Louis R. Kurzmiller
Jonathon Kutler
Huang Kwei
Alton Lacks
Carol J. Lacombe
James Lacombe
Brook Lakew
John Lallande
Allan C. Lane
Paul Lang
James Lanks
Annette LaJoie
Rhea Langlois
Douglass J. Lankenau
James G. Lanks
James Largent
Gerald Larner
Jeanette Larsen
Kimberley Lascola
Warren C. Lathe, Jr.

Connie Lau
Robert E. Laughlin
Kenneth W. Laverly
C. Laverty
Santo Lavorata
Frank Lawrence
John Lawrence
Robert L. Leavel
Deborah Lebair
Robert S. Lebair, Jr.
Javier Lecha
Maria Lecha
Elwin Lee
Grace Lee
Norman Lee
Roger Lee
Oleina Lee-Simpson
Henning Leidecker
Joel Leifer
Roland F. Leland
Darrel D. Lemon
Margie D. Lenson
Michael Lenz
Theodore Leoutsakos
Ralph Leshun
Stephen O. Leslie
Jim M. Lester
Brian Lev
Steven Levin
Allen J. Levine
Douglas B. Leviton
Edward P. Lewis
William Lewis
Robert Light
Michele Limon
Michael Lin
Celeste Lindsay
Raymond Lindsay
Jerome Lindsey
John Lindsey
Charles Lineweaver
Loren Linstrom
Moussa Lishaa
Casey Lisse
Donald E. Lloyd
Patrick K. Lo
Charles David Lockhart
Karen Loewenstein
Barry R. Lohr
Catherine A. Long
Lois Long
Gerald Longanecker
Manuel P. Lopez
Blake Lorenz
John Lorenz
David L. Love
Brian Lowe
Crockett Lowe
Barbara Lowery
Paul Loyd
Philip M. Lubin
Ray A. Lundquist
Lorraine Lust
Ronald A. Luzier
Michael Lynch
Gene McAlicher
Marsha McBride
Dennis McCarthy
Patrick McCaslin
Marty McClure
Sidney McClure
Matthew McCoy
Duane K. McDermond
Buck McDonald
Darrell McDonald
Deborah McDonald
Gary A. McDonald
James McDonald
Mickey McDonald
Renee McDonald
Wesley McDonald
William R. McDonald
Olivia V. McFarlane
William McGunigal
Daniel C. McHugh
Walter McKenny
Christopher McLeod
John McMahon
Joyce McMurtrey
Vernon B. McNeill

Snehavadan Macwan
John McWilliams
Sean Madine
Laurence E. Madison
Timothy J. Madison
Edward Magette
Thomas Magner
Jeffer M. Mahmot
Robert Maichle
David Maitland
John A. Majane
Angela E. Majstorovic
Andrew Makar
James A. Mallin
William A. Mamakos
George Mangum
Kevin Mangum
John Mangus
Amil Mann
Mark Mann
Roland R. Manning
Dominic Manzer
Daniel Marinelli
Paul Marionni
Jane Marquart
Thomas H. Marsh
Anthony Marshall
Dale Marshall
Earl Martin
Frank Martin
Pilar Martin
Robert Martin
Gregory Martins
John Maruschak
John Marvin
Clair L. Maschauer
Shirley Masiee
James Mason
Derck Massa
John C. Mather
Eric Mathis
Philip Matthews
Andrew Mattie
Carmen Mattiello
Kelly Mattis
Roger Mattson
Henry Maurer
Mary S. Maxwell
L. Eliezer May
Houston Mayhall
Riley Mayhall
Frank Mayo
James O. Mayor
George Meadows
Robert D. Medlock
David Medwedeff
Cindy A. Meisner
John Mejane
Nicolas V. Mejia
Fred C. Menage, Jr.
John Mengel
Robert Menrad
Maureen Menton
Sandra L. Mercil
Darrell Merrill
James Metz
Harold Meyer
Stephan S. Meyer
William H. Meyer
Joseph Miko
Laura Milam
Malcolm B. Milam
Andrew Miller
Anthony Miller
Barry Miller
Curtis Miller
Grace Miller
Henry Miller
Kelly A. Miller
Mark Miller
Paul Miller
William Miller
James Milligan
James Mills
Marilyn Mills
Richard C. Mills
Thomas Mills
Paul Milwicz
Linda Miner
James Ming

Peter Minott
William Mish
Darryl Mitchell
Joel Mitchell
Kenneth Mitchell
Reginald S. Mitchell
Mila Mitra
Faye Mitschang
Herbert J. Mittleman
William Mocarsky
Gary Moffatt
David Mok
Stephen Monk
Clarence R. Monn, Jr.
Patrick Moody
Keith Moore
John Mooring
Armando Morell
Barbara Moroz
Carolyn Morris
Richard Morris
Samuel H. Moseley, Jr.
Carol L. Mosier
Mahmoud M. Motamedamian
Doug H. Mount
Charles Moutoux
Faye Movahhed
Paul Mowatt
Ronald C. Moxley
Earl Moyer
Donald C. Muganda
Joseph Muller
Ronald M. Muller
James Mullins
Jeggery Mullins
Michael Mumma
James Munford
Joseph A. Munoz
Alan Murdock
Thomas L. Murdock
Kenneth Murphy
Paul Murray
William Myers
David Nace
Edward Nace
Jolyn Nace
Roy Nakatsuka
Cookie Namkung
Richard Nash
Elaine Nelson
Norman Ness
Hellen Neumann
Douglas Newlon
Phillip Newman
Son Ngo
Tu Nguyen
Lawrence Nichols
Harry Nikirk
David Niver
Peter Noerdlinger
Louis M. Norman
Linda Norsworthy
Marci Norton
Ramon T. Novo
Prasad Nune
Joseph O'Connor
Nils Odegard
Sten Odenwald
Ted Ogelsby
John H. Olesen
Leonard Olson
Terence P. O'Neill
Matthew A. Opeka
David Orbock
Richard L. Ore
Becky M. Ortega
Edmond Oser
West Over
Linda Owens-Ryschkewitsch
Larry Pack
John Packard
Frank Paczkowski
Stephen J. Paddack
Nancy Painter
David Palace
Mark Palini
Aliza Panitz
Bradford H. Parker
Mark Parker
Brenda Parkinson

Jeta C. Parraway
Lillian Parraway
Ziba Parsa
John Parsons
Herb Pascalar
Suryakant M. Patel
Lilian Patrick
Robert W. Patschke
Frederick Patt
George J. Pattison
Nancy Patton
Steven Patton
John Paulkovich
Dan A. Payne
Jack Peddicord
Jeffrey Pedelty
David Perretin
Freddie L. Perry
James C. Perry
William Perry
George Peters
Robert L. Peterson
Steven Peterson
Yvonne Peterson
David C. Pfenning
Dung Pham
Minh C. Phan
James E. Phenix
John Phillips
Robert Phillips
George Pieper
Brian Pierman
Fred Pirnia
Ryszard L. Pisarski
Samuel Placanica
Barbara Plante
Juergen P. Pleines
Walter Plesniak
John Pocius
James Poland
Darrel J. Poloway
Karen Pope
Bradley J. Poston
Donald N. Potter
Angela Powell
Bernard Powell
Carl Powell
Gwendolyn Powell
Charles Powers
Edward I. Powers
Michael K. Powers
Urmila Prasad
Quain S. Prather
Lawrence A. Pratt
Henry Price
Robert Price
Phillip C. Prichard
Theresa Proctor
Harper Pryor
David Puckett
John Pyle
Chadwick T. Quach
Edward R. Quinn, Jr.
Harvey Rabinowitz
William Raborg
Marie Rabyor
Atul Rae
Paul Ragan
Rebecca Ragusa
H. K. Ramapriyan
Rajendra G. Ramlagan
Ramjit Ramrattan
Abraham Ramsey
William Rappel
Roger Ratliff
Thomas Ratliff
Lowell Rau
Anne Raugh
Gayle Rawley
Shirley Read
Ronald C. Rector
Kevin Redman
Yvonne Denise Reed
William J. Regan
Jim K. Reidy
Leshun Relph
Albert Renner
Bruce W. Rentch
Kirk Rhee
Catherine Richards

Donald Richardson
George Richardson
Carl Riffe
Carl Riley
Nancy Rinker
Wyatt Rinker
Katharyn Rivas
Michael Roberto
Dwight Roberts
Joan Roberts
Deloris Robinson
George Robinson
James Robinson
Jerry Robinson
Jon W. Robinson
Vernon Robinson
Willie G. Robinson, Jr.
Jill Rock
Edward Rodriguez
Guillermo E. Rodriguez
Otilia I. Rodriguez-Alvarez
Thomas Roegner
James S. Rogers
Lonnie Joe Rogers
Laurie Rokke
Kent G. Roller
Richard R. Rolnicki
Arlett Romano
Juan Roman-Velazquez
Anthony Rondeau
Francis A. Rondeau
Giulio Rosanova
Lawrence Rosen
Jacob Rosenberg
Robert Rosenberry
Charles Ross
Peter Rossoni
Anthony Rota
Charles Roten
Betty Rowdon
Shelley Rowton
Mary Royce
Laddawan Ruamsuwan
Gregory A. Ruck
Janet K. Ruff
Arthur P. Ruitberg
George Rumney
Richard Runyon
Shirley Rupp
Thomas Russell
P. Russo
Richard C. Rutkowski
Timothy Rykowski
Michael Ryschkewitsch
John A. Ryskewich
Jack Saba
Debora B. Sabatino
William Sadler
S.R.K. Vidya Sagar
Joseph W. St. John
Andrew G. Samchuck
Henry Sampler
Robert Sanchez
Edward Sanford
Robert Sanford
Stephanie M. Sanidas
Jairo Santana
Jay Santry
Brett P. Sapper
Carol Sappinton
Mark Saulino
Dwayne A. Saunders
Rudolfo Scarpati
Clell Scearce
Charles Scharmann
John J. Schaus
George C. Scheffey
John Scheifele
William Schiflett
Richard Schmadebeck
Gregory Schmidt
Josef A. Schmidt
Sherry Schmitz
Michael Schools
Robert T. Schools
Kenneth B. Schou
Michael Schroedl
Evelyn Schronce
Diane Schuster
David Schwartz

D. P. Schwartz
Ronald Schweiger
John Scialdone
Courtney Scott
Jill Scott
Richard H. Scott
Sidner J. Scott
Kay Scoville
Stanford Scribner
Lawrence Sears
Diane Sebok
Lou Segar
Ritchie Seigel
Paul Seiler
Roman I. Semkiw
Steven Senk
Ignatius N. Serafino
Aristides T. Serlemitsos
Debra Servin-Leete
Steven Servin-Leete
Eugene V. Shackelford
Sharon Shackleford
Richard A. Shafer
Edward Shaffer
Kamlesh Shah
Michael C. Shai
Rakeev K. Sharma
John Shawhan
Oren Sheinman
Charles Sheppard
Allan Sherman
T. R. Shevlin
William Shields
David Shrewsberry
Frederick Shuman
Charles Shyr
Christopher J. Silva
Allan Silver
Robert F. Silverberg
Joyce Simcox
Robert Simenauer
Mornam F. Simms
Carl G. Simon
Stephen Simonds
Alda Simpson
Douglas Sims
Michael Singer
Kawal P. Singh
Ramesh Sinha
John Arthur Skard
Joseph T. Skladany
Edward Skolka
Rosalie Skrabak
Shereese Sloan-Smith
Marjorie A. Small
Earnestine Smart
Jan Smid
Charlene Smith
Frank Smith
Frederick M. Smith
George Smith
James Smith
Larry L. Smith
Lee Smith
Margaret Smith
William Smith
George F. Smoot
Edward Smygocki
Thomas Sodroski
Carl Solomon
Harry Solomon
Kyung-Hye Song
William Spiesman
Kathryn Spinola
Michael C. Spinolo
George R. Springham
Walter Squillari
William Stabnow
Carolyn H. Stahl-Boss
Kenneth Stark
Richard A. Stavley
Joseph Stecher
Sally Stemwedel
Joseph Stephenson
Robert Steudel
Peggy Stevens
David Stevens-Rayburn
Alphonso Stewart
John Stewart
Kenneth P. Stewart II

George Stitt
Joseph Stivaletti
Neal D. Stoffer
Jonathan Stokes
Edward J. Stoner, Jr.
George Stoner
Kathryn Stoner-Jhabvala
Cedley Storey
Allan Strojny
Priscilla Struthers
Mary Jean Studer
Philip Studer
Douglas D. Stump
John Sudey
Bobby A. Sullivan
David J. Sullivan
Edward Sullivan
Thomas Sullivan
Janet Sushon
Steven G. Sutherlin
Betsy Sutton
John Sutton
Robert Sutton
Thomas L. Swales
Maximilian Swantko
John Sween
Robert Swope
Carl Sylvis
Kasimer J. Szczech
Wiktor Szczyrba
Darnell Tabb
Devin Tailor
John J. Tarpley
Cynthia Tart
Christopher M. Tasevoli
Charlene Taylor
Gerald Taylor
Karen Taylor
Muriel Taylor
Luis Tenorio
William Tereniak
Krishna Tewari
Janet Tharps
Khon Thia
Charles Thomas
William Thompson
John Thomson
Richard Tighe
Smith E. Tiller
Gary Toller
John Tominovich
Hollice Tommer
Raymond W. Topolski
Marco Toral
Hank Torrance
E. Torres
Sergio Torres
Eduardo Torres-Martinez
Gregory Toth
John Townsend
Harrison W. Toy
Paul Trahan
Alice Trenholme
Robert Trescott
Jack Triolo
Wen-Hu Tsai
Jim L. Tucker
Angela Tung
Kevin Turpie
Joseph Turtil
Timothy Tyle
Nigel Tzeng
Joanne Uber
Ralph Uline
Bernard Uphsur
A. Robert Urbach
Danny J. Utley
Sandra Vanderweit
Don A. Van Gundy
John T. Van Sant
Michael Van Steenberg
Jack Van Zant
Barbara Vargo
George E. Varndell
Frank Varosi
Edwin Vaughan
Gilbert Velasco
Thomas J. Venator
Ravi Venkataraman
Ralph Vidnovic

Jancilon Viegas
Michael Viens
Milton Vineyard
Michael Vinton
David W. Voight
Daniel Volonakis
Eugene Volpe
Stephen M. Volz
Robert G. Voorhees
Clarence Wade
Carl Walch
James Wall
Nancy Walpole
Mark Walter
Kenneth Walters
David Wampler
Harte Wang
Dalroy Ward
Douglas Ward
John Ward
Rodney Ward
Shari Warner
Timothy Warner
David K. Wasson
Donald Watson
Sharon Watson
Robert C. Weaver
John Webb
Richard R. Weber
Brooke Webster
Timothy Wechsler
Janet Weiland
Ronald C. Weimer
Howard Weiss
Rainer Weiss
Robert Welbourne
Dott Wells
Jerry A. Wells
Robert F. Wendler
Albert West
Donald West
Steven M. West
Mark Westerman
David Wheeler
Benjamine White
Carrie J. White
Charles A. White
Debra White
Richard White
Rickey L. White
Steven White
Raymond Whitehead
Raymond Whitley
Edward Wienike
Sandra Wigton
David T. Wilkinson
Bernadette Williams
Billy B. Williams
Domenica Williams
Herman Williams
Michael Williams
Seth Williams
Steven Williams
Clayton Wilson
Denise Wilson
Edward Wilson
John Wilson
Lisa Wilson
Meridith Wilson
Reid Wilson
Richard Wilson
Thomas L. Wilson
Robert T. Wimsatt
Julia Winkler
Allan N. Winters
Howard Witcher
Christopher Witebsky
Berl Wittig
Carl Wittman
John Wolfe
Robert Wolfe
John Wolfgang
Gary Wolford
Jack Wolsh
James Wood
Roger Wood
Rose Wood
Steven Wood
Christopher Woodhouse
Natalie Woodson

William Woodyear
Lois Workman
Everett Worley
Louis Worrel
Barbara Wrathall
Edward L. Wright
Leroy Wright
Robert Wright
J. Andrzej Wrotniak
David Wyckoff
David Wynne
Michael Yachmetz
John Yagelowich
John H. Yamada
Jack Yambor
Jaya Yodh
David D. Yoest
Song Yom
Suk J. Yoon
Anthony Young
Earle Young
Kenneth Young
Kimberly Young
Peter Young
Stephania B. Young
Wesley Young
John L. Zahniser
John Zaniewski
Jane Zellison
Gerald R. Zgonc
David J. Zillig
Jerry Zimmerman
Lynne Zink

COMPANIES

AeroJet General
Applied Research Corporation
AT&T Information Services
Ball Aerospace Systems Division
Bendix Field Engineering Corporation
Berkshire Technologies
City Wide Security Services, Inc.
Computer Based Systems, Inc.
Computer Sciences Corporation
Electromagnetic Sciences, Inc.
Electronic Associates, Inc.
Engineering & Economic Research
Fairchild Space Company
Ford Aerospace Division
General Electric Corporation
General Sciences Corporation
Hughes Aircraft Company
Hughes STX Systems Corporation
Information Development & Applications, Inc.
International Business Machines
J. H. Lawrence Corporation
McDonnell Douglas, Inc.
NSI Technical Services Corporation
NYMA, Inc.
OAO Corporation
RMS Technologies, Inc.
Science Application Research
Science Systems Applications, Inc.
Stanford Telecommunications
Swales & Associated, Inc.
UNISYS
Universities Space Research Association

To Dig Deeper: Further Readings

Barrow, John, and Joseph Silk. *The Left Hand of Creation: The Origin and Evolution of the Expanding Universe*. New York: Basic Books, 1983.

Cornell, James. *Bubbles, Voids, and Bumps in Time: The New Cosmology*. New York: Cambridge University Press, 1989.

Davies, Paul. *The Cosmic Blueprint*. New York: Simon & Schuster, 1988.

———*The New Physics*. Cambridge University Press, 1987.

———*Coming of Age in the Milky Way*. New York: William Morrow, 1988.

Ferris, Timothy. *The Red Limit*. New York: William Morrow, 1977.

Gamow, George. *The Creation of the Universe*. New York: Viking, 1957, of historical interest.

———*Mr. Tompkins in Paperback*. New York: Cambridge University Press, 1965.

Gribbin, John. *In Search of the Edge of Time*. London: 1992.

———*The Omega Point: The Search for the Missing Mass and the Ultimate Fate of the Universe*. New York: Bantam, 1988.

Guth, Alan, and Paul Steinhardt. "The Inflationary Universe," *Scientific American*, January 1990, pp. 116–128.

Harrison, Edward. *Darkness at Night: A Riddle of the Universe*. New York: Cambridge University Press, 1987.

Hawking, Stephen. *A Brief History of Time*. New York: Bantam, 1988.

Krauss, Lawrence M. *The Fifth Essence: The Search for Dark Matter in the Universe*. New York: Basic Books, 1989.

Layzer, David. *Cosmogenesis: The Growth of Order in the Universe*. New York: Oxford University Press, 1990.

Lederman, Leon, and David Schramm. *From Quarks to the Cosmos.* New York: Scientific American Library, 1991.

Lightman, Alan P. *Ancient Light: Our Changing View of the Universe.* Cambridge, Mass.: Harvard University Press, 1991.

Longair, Malcolm. *The Origins of Our Universe.* New York: Cambridge University Press, 1991.

Overbye, Dennis. *Lonely Hearts of the Cosmos: The Story of the Scientific Quest for the Secret of the Universe.* New York: HarperCollins, 1991.

Parker, Barry. *Invisible Matter and the Fate of the Universe.* New York: Plenum Press, 1989.

Reeves, Hubert. *Atoms of Silence.* Cambridge, Mass.: MIT Press, 1981.

———*L'heure de s'enivirer: L'univers a-t-il sens?* Paris: 1986.

———*Patience dans l'azure: L'évolution cosmique.* Paris: 1981.

Riordan, Michael, and David Schramm. *The Shadows of Creation: Dark Matter and the Structure of the Universe.* New York: W. H. Freeman & Co., 1991.

Schilling, Govert. *De Salon van God: Spuertocht naar de Architectuur van de Kosmos.* Amsterdam: Wereldbibliotheek, 1993.

Silk, Joseph. *The Big Bang.* New York: W. H. Freeman & Co., 1989.

Trefil, James. *The Dark Side of the Universe.* New York: Doubleday, 1989.

——— *The Moment of Creation: Big Bang Physics from Before the First Millisecond to the Present Universe.* New York: Macmillan, 1984.

Weinberg, Steven. *The First Three Minutes: A Modern View of the Origin of the Universe.* New York: Basic Books, 1977.

Alternative Cosmologies Books:

Alfven, Hannes. *Worlds-Antiworlds.* San Francisco: W. H. Freeman, 1966.

Arp, Halton C. *Quasars, Redshifts, and Controversies.* Berkeley, Calif.: Interstellar Media, 1987.

Lemaître, Georges. *The Primeval Atom.* New York: Van Nostrand, 1950.

Learner, Eric. *The Big Bang Never Happened: A Startling Refutation of the Dominant Theory of the Origin of the Universe.* New York: 1991.

Acknowledgments

The awesome mysteries of the universe are alluring to us all. This book attempts to bring to the public some of the sense of wonder and excitement that researchers experience. It also acknowledges the debt we have to the public, who through their taxes supply the resources that make our work possible. I thank the U.S. agencies—NASA, the Department of Energy, and the National Science Foundation—which supported this work in many ways but particularly through their bases and laboratories such as the Lawrence Berkeley Laboratory, the Space Sciences Laboratory, the National Scientific Balloon Facility, the South Pole Station, NASA Ames, Goddard, and other space-flight centers, and the many other observatories and scientific centers. The staff in these places have been of the highest caliber and most dedicated in their efforts. It has been a real pleasure and privilege to work with them. I especially thank the heads of the LBL Physics Division—Bob Birge, Dave Jackson, Pier Oddone, and Bob Cahn—and the directors of the Space Sciences Laboratory—Kinsey Anderson, Buford Price, and Chris McKee—for their support and for providing very exciting, creative, and productive research facilities.

All of this support and infrastructure would not be fruitful were it not for a strong educational system. I gratefully acknowledge my teachers, beginning from the earliest times up through the university level. At every level I was given the opportunity and encouragement to learn and to study the work of previous generations. My teachers instilled in me a respect for knowledge, and books (I still can't bear to part with one), and a joy and curiosity to learn. By the time I got to college my professors had an easier task—that of funneling to me the vast knowledge already accumulated by previous scholars and encouraging me to seek my own interests and levels. I thank the leaders of MIT's educational process—Victor Weisskopf, Philip Morrison, Steven Weinberg, and many more. Special acknowledgment goes to my thesis adviser, David H. Frisch, who pulled and pushed me into experimental physics and trained me in those aspects that are not found in textbooks, and to Luis W. Alvarez, who continued that training at Berkeley and encouraged me to expand my work. The excellence of the country's universities and the continuing education and training they provide is one of our greatest assets.

Naturally everything begins at home and I must thank my family and friends for their encouragement and continuing support. I especially acknowledge the role of my mother, who provided extensive encouragement and an excellent example as a person and as a science teacher. She instilled in me a love and respect for culture and learning. My father was a role model by action and deed—an explorer and adventurer in the transition from the old traditions to the new scientific searchers. My friends were understanding and cheerfully put up with my long hours and absences in the name of art and science, especially Julian and Connie.

Research in cosmology and science in general is a large-scale effort involving many and varied people. I and the rest of us owe a great debt to those who devote their efforts to this enterprise. Their work makes us all richer in many ways. The story in this book results from the hard-won efforts of very many. It has been my pleasure to work with many—including the prominent scientists the public has heard about as well as less well-known researchers. Also I worked with the engineers and technical staff who are rarely known but vital nevertheless. At Berkeley it has been my pleasure and benefit to work with many such excellent people, including Jon Aymon, Hal Dougherty, John Gibson, Doug Heine, Robbie Smits, and John Yamada. I also worked with many outstanding scientific colleagues—undergraduate students, workhorse graduate students, post docs, and more senior staff.

The COBE project involved coworkers and colleagues on a truly massive scale. The Appendix is a list of more than a thousand people who contributed to COBE. Special acknowledgment goes to the COBE science working group—Chuck Bennett, Nancy Boggess, Ed Cheng, Eli Dwek, Sam Gulkis, Mike Hauser, Mike Janssen, Tom Kelsall, Phil Lubin, John Mather, Steve Meyer, Harvey Moseley, Tom Murdock, Rick Schafer, Bob Silverberg, Rainer Weiss, Dave Wilkinson, and Ned Wright—and to the various members of the DMR team whose composition changed over the year as the experimental phases changed. More than forty people contributed in a major way to the DMR and they deserve special credit. However, so do all those who made it possible for the DMR (and FIRAS and DIRBE) to achieve their goals and thus the Appendix acknowledgement.

Finally, I would like to thank all those involved with this book, including Keay Davidson, Roger Lewin, William Morrow personnel—editor Maria Guarnaschelli, Chas Edwards, Arlene Goldberg, Kurt Aldag, and the Production Department—as well as our other publishers and our agent, John Brockman. Christopher Slye took my simple sketches and concepts and made many of the beautiful illustrations that add so much to this book. I would also like to thank all those who read and commented on various sections of the book. Most of all, thanks to the readers who I hope enjoyed and learned from this book. They represent the informed public and those who support science and culture.

Index